PRAISE FOR ANNIE DILLARD

"Annie Dillard is a writer of unusual range, generosity, and ambition. . . . Her prose . . . is bracingly intelligent, lovely, and human."
—*Boston Globe*

"There is none better than . . . the great Annie Dillard, an essayist of savvy, wit, and uncanny insight."
—Kevin Barry

"[One of] those who first made me want to try writing essays and set the standard to reach for when I tried, even as their more abiding commitment to the form keeps me humble by comparison."
—Jonathan Lethem

"Annie Dillard is, was, and always will be the very best at describing the landscapes in which we find ourselves."
—*Star Tribune* (Minneapolis)

THE ABUNDANCE

ALSO BY ANNIE DILLARD

General Nonfiction

Pilgrim at Tinker Creek

FOR THE TIME BEING

Holy the Firm

The Writing Life

Memoir

An American Childhood

Fiction

The Living

The Maytrees

Journalism

Encounters with Chinese Writers

Theory

Living by Fiction

Poems

Tickets for a Prayer Wheel

Mornings Like This

THE ABUNDANCE

NARRATIVE ESSAYS OLD AND NEW

ANNIE DILLARD

Foreword by Geoff Dyer

An Imprint of HarperCollins*Publishers*

HarperCollins books may be purchased for educational, business, or sales promotional use. For information please e-mail the Special Markets Department at SPsales@harpercollins.com.

A hardcover edition of this book was published in 2016 by Ecco, an imprint of HarperCollins Publishers.

FIRST ECCO PAPERBACK EDITION PUBLISHED 2017.

Designed by Sunil Manchikanti

Library of Congress Cataloging-in-Publication Data has been applied for.

ISBN 978-0-06-243296-4

19 20 OV/LSC 10 9 8 7 6 5

For Phyllis Rose

They will question thee concerning what they should expend.
Say: The Abundance.

—THE QUR'AN, SURA OF THE COW

CONTENTS

CONTENTS

FOREWORD

A passage from *An American Childhood*—presented here as a short essay called "Waking Up"—ends with the characteristically brilliant image of a woman diving into water, becoming sealed in her reflection, and wearing it "as she climbs rising from the pool, and ever after." It's Annie Dillard all over, that passage, that image. Waking up (coming into consciousness), remaining wide awake, leading "a life of concentration," rather than sleep-wading through life, have been her abiding concerns. Her first book, *Pilgrim at Tinker Creek,* begins with the author waking "in daylight to find my body covered with paw prints in blood."

A writer who never seems tired, who has never plodded her way through a page or sentence, Dillard can only be enjoyed by a wide-awake reader. By which I really mean that she rejuvenates even the word-weary, who—it can happen to the best of us—have succumbed, like Henry in John Berryman's fourteenth *Dream Song,* to being bored by literature, "especially great literature." Maybe that's part of her appeal: the way that her best work often comes in shapes and forms unencumbered by the protocols associated with "great literature." Some of the pieces here were first collected in *Teaching a Stone to Talk,* which may have looked, when it came out, like the kind of interim report "a writer brings out to supplement his real work." But no, she explains in the author's note, "this is my real work, such as it is."

Genre-resistant nonfiction may be a recognized genre these days, but Dillard awoke to its possibilities—and attendant difficulties—back in the early 1970s, when she was writing what became *Pilgrim at Tinker Creek.* "After all," she noted in a journal, "we've had the non-fiction novel—it's time for the novelized book of nonfiction." Easier said than done, of course. As she puts it in *The Writing Life,* "Writing every book, the writer must solve two problems: Can it be done? And, Can *I* do it?" But it's the nature of this beast that the answer to the second question becomes known only after the book has been written—or has failed to get written—when there's no longer any point asking it. This is different from geographical exploration and discovery, where, once you get to the place to be discovered—there it is, waiting to be claimed and named!—the uncertainty is over. The discovery in art is often gradual, a process of minor discoveries riddled with uncertainties and the potential for making that which is discovered vanish before your eyes, like a mirage.

"What kind of book is this?" Dillard asked herself of *Pilgrim at Tinker Creek,* a question that will continue to be asked for as long as the book is read. We read it, in part, to find the answer and, after we have read it—this is the great thing—are still not sure.

So, one question giving rise to another: What kind of writer is Annie Dillard? One of those writers, she decided early on, who chose to define herself in extra-literary terms: "an explorer of the neighborhood," "a fugitive and a vagabond, a sojourner seeking signs," and "a stalker, or the instrument of the hunt itself." This last idea is especially telling. The Stalker in Tarkovsky's eponymous film says of a teacher (called Porcupine, as it happens) that "he opened my eyes." Dillard opens our eyes to the world and to new ways of articulating what we see. More exactly, she alerts us to the possibility of being free of formal conventions that constrain in the guise of enabling. She does this when you first read her and does so again when you reread her. She did it in *Pilgrim at Tinker Creek,* and she is doing it again here, in what might be regarded as a reframing or rehang of much-loved work. The pleasure and excitement are intense, whether you know the original material well or are encountering it—waking to its astonishments—for the first time.

After that fanfare, let's deflate things a bit by introducing a writer who must surely be considered as an antithetical rather than kindred spirit. E. M. Cioran, author of *The Trouble with Being Born,* claimed to owe everything, torments and insights alike, to his insomnia. If Dillard celebrates the day ("Every day is a god, each day is a god"), Cioran was condemned to wakefulness, to spending his nights wandering in a blaze of unrelieved consciousness. Yet a strange negative affinity exists between the two. They are, for example—though it is not necessarily the first thing that strikes you about either—wonderfully comic writers. In Cioran the comedy is funereal to the point of absurdity (which *is* the point); in Dillard it's the comedy of rapture. Or at least it's a comedy that permits prose and thought to soar while inoculating the rapturous against the three ills of which nature writers should live in permanent dread: preciousness, reverence, and earnestness (the sure sign, Nietzsche reminds us, of a slow brain). "The mind wants the world to return its love, or its awareness; the mind wants to know all the world, and all eternity, even God," she writes in the classic essay "Total Eclipse." And then continues: "The mind's sidekick, however, will settle for two eggs over easy."

On the humor front it helps, also, that Dillard's pretty

much a fruitcake. In an insightful review of *Pilgrim at Tinker Creek,* Eudora Welty confessed that at certain points "I honestly do not know what she is talking about." And *Pilgrim* is far from being Dillard's most difficult or nuttiest offering; that honor would have to go to *Holy the Firm* (really bonkers, and all the more enjoyable for it). Incomprehension is usually the result of obfuscation, the words refusing to slip into focus; Dillard, however, remains a writer of exceptional clarity, even when we are struggling to grasp the meaning of what is being said so clearly, so *brightly*.

Perhaps it's this very lucidity that compels her to peer into the darkness of ethical and metaphysical questions. She has returned frequently to the old chestnut of suffering in the world ("Cruelty," she writes in an insight worthy of Simone Weil, "is a mystery, and the waste of pain"), and it is here—to go back a paragraph—that we find more common ground with Cioran. In his despair as a young man, Cioran decided that "philosophy is no help at all, and offers absolutely no answers. So I turned to poetry and literature, where I found no answers either, but states of mind analogous to my own." Dillard's position on these matters is set out by a character in her novel *The Maytrees:*

> Having limited philosophy's objects to certainties, Wittgenstein later realized he broke, in however true a cause,

> his favorite toy, metaphysics, by forbidding it to enter anywhere interesting. For the balance of Wittgenstein's life he studied, of all things, religions. Philosophy . . . had trivialized itself right out of the ball park. Nothing rose to plug the gap, to address what some called "ultimate concerns," unless you count the arts, the arts that lacked both epistemological methods and accountability, and that drew nutty people, or drove them nuts.

These "ultimate concerns" are boiled down, on several occasions in Dillard's work, to a single nutty question: "to wit, What in the Sam Hill is going on here?" Though she has written two fine novels, her answers tend not to come in novel form. The novel, after all, is generally programmed to describe and map the social landscape. This is neither her main area of thematic interest nor where her talents are uniquely displayed. Early on in *An American Childhood,* young Annie discovers "that I myself was both observer and observable, and so a possible object of my own humming awareness." Welty, in her review of *Pilgrim,* was less than wholehearted in her admiration for this canary-in-the-me-mine method: "Annie Dillard is the only person in her book, substantially the only one in her world; I recall no outside human speech coming to break the long soliloquy of the author." If this, for Welty, was almost the opposite of what it meant to be a writer, it was exactly what attracted Cioran to mystics and saints. Hostile to religion, he nevertheless developed an interest in the mystics because they "lived a more intense life than others. And, too, because of their extraordinary pride, me and God, God and me." It is here, amidst what she calls "the literature of illumination," that Dillard declares her genius most openly. Actually, let's put that differently: It's here that she has made her home.

Bear in mind also—the antecedents of Emerson and Thoreau are important—that while the light may be universal, it always falls in a particular way on a particular patch of ground. "I never saw a tree," Dillard declares, in a valuable piece of advice to writers of any and every stripe, "that was no tree in particular." At one point she insists, with the unflappable resolve of a lawyer's dream witness, on having seen an angel in a field. I'm of the Courbet persuasion but am inclined to believe her, less because she conceived of *Pilgrim at Tinker Creek* as "what Thoreau called 'a meteorological journal of the mind' " than because, after careful deliberation, she confided to her journal that *Walden* was "really a book about a pond." In everything she writes she subscribes to the idea—attributed variously to Éluard or Yeats—that "there is another world, but it's in this world." Wherever you are in this world, she reminds us, "life is always and necessarily lived in detail."

Inevitably, then, she has been scrupulous in detailing her whereabouts: where she was from (*An American Childhood*), where she was living (*Pilgrim at Tinker Creek*), and where she was going (*Teaching a Stone to Talk*). That kind of attentiveness means that she also needs to factor in whichever books she has in her duffel bag to sustain and nourish her in these places and phases. And so, at the risk of uprooting her from the native soil of Emerson and Thoreau, I feel similarly at liberty to open up my own bag of books and mention the three writers who, as the saying goes, beat a path to her door for me.

In *Emerson: The Mind on Fire,* Robert D. Richardson (Dillard's husband) enables us to see—almost to share—how Emerson became the writer he did by reading the writers he read. These are the writers who formed him. But what of the path that draws readers to a writer they come to love? That, too, is formed by writers who not only prepare us for such encounters, but also subtly prepare the writer *for us*. In turn, this writer will prepare the way for other encounters further along the road. For the time being, though, Dillard is where I'm at: a temporary terminus.

The three writers who got me to this point have much in common with Dillard, not least a tendency—often regarded as fatal flaw in imaginative writers—to offer advice: one reason, perhaps, why none of them could confine their talents to the novel. First there is D. H. Lawrence, one of the most insistent and vehement offerers of advice in the English literary canon. But Lawrence's hieratic tendency, the unwavering belief that he had some kind of cure for the sickness of his times, is always predicated on "his relationship, his bond with everything in creation." The words are those of his widow, Frieda, who continued to be amazed by this—"no preconceived ideas, just a meeting between him and a creature, a tree, a cloud, anything"—long after his Salvator Mundi sermonizing had become wearisome. We are in the jurisdiction of the detail, so let's take a tiny example. When Lawrence writes of cypress trees in *Twilight in Italy,* "For as we have candles to light the darkness of night, so the cypresses are candles to keep the darkness aflame in the full sunshine," he not only makes us see these particular trees in a very particular light; he also prepares our eyes for seeing (reading) pages by Dillard, which, in turn, bring Lawrence flaring back into life as her—and our—permanent contemporary.

Then there is Rebecca West. The fact that her tone and register are so different—grand where Dillard is scatty—should not distract us from the similarities both incidental and overarching (and therefore, ultimately, tonal). By baking a cake for friends, West realizes in *Black Lamb and Grey Falcon,* "one is striking a low note on a scale that is struck higher up by Beethoven and Mozart." In *The Writing Life,* Dillard expresses her admiration for a stunt pilot. "It was as if Mozart could move his body through his notes, and you could walk out on the porch, look up, and see him in periwig and breeches, flying around in the sky. You could hear the music as he dove through it; it streamed after him like a contrail."

More generally, there is the shared tendency, the strength of the urge, towards ultimate concerns. Asked to write a book about empire, West declined because she had nothing to say "except for the fancy bits on religion and metaphysics that I would throw in in my demented way." In *her* demented way, Dillard is content to throw a lot overboard to make room for the metaphysics. But there is nothing desperate, anorexic, or body-denying about her preoccupation with questions of the spirit. In the concluding essay here the quest for the absolute is represented both by a religious ceremony and by an expedition to the North Pole.

Her idea of meta is unashamedly physical. "What's wrong with golf?" she asks in "This Is the Life." "Nothing at all." Well, there's quite a lot wrong with it, in my view; mainly that it's not tennis or volleyball. And it is tennis, rather than golf, to which she turns when urging writers to "hit the edges." I once heard tennis coach Brad Gilbert give similar instructions to one of his protégés. The phrase he shouted out was "Stretch the court!" but that's another story. Or maybe it is precisely the story. See, you can stretch the court by stripping it of a lot of the clutter required by the novel—all the furniture of character—but the physicality of storytelling must remain strong; must be doubly strong, in fact.

In *And Our Faces, My Heart, Brief as Photos,* John Berger—the third point in the British triangle that enabled me, in a most un-Bermuda-like way, to find Dillard—writes that "the traffic between story-telling and metaphysics is continuous." Here's a little example of that (foot) traffic from *Teaching a Stone to Talk:*

> You know what it is to open up a cottage. You barge in with your box of groceries and your duffel bag full of books. You drop them on a counter and rush to the far window to look out. I would say that coming into a cottage is like being born, except we do not come into the world with a box of groceries and a duffel bag full of books—unless you want to take these as metonymic symbols for culture. Opening up a summer cottage is like being born in this way: at the moment you enter, you have all the time you are ever going to have.

Almost out of time myself, I can only add that this last is from an essay, "Aces and Eights," that didn't make it through the door of the present volume, a rigorously curated collection if ever there was one. It's a reminder of the abundance of material waiting outside, in the rest of her books. You can head over there later.

For now: Welcome, step inside.

GEOFF DYER

ACKNOWLEDGMENTS

Special thanks to: Ann Beattie, Marc Chenetier, Cody Rose Clevidence, Hal Crowther, Ophelia Dahl, R.H.W. Dillard, Wendy Doniger, Paul Farmer, Amy Fields, Dan Halpern, Bridget Read, Edward Hoagland, the late Gary Isaacson, Lewis Lapham, Gabi Marcus, Rob McQuilkin, Maggie Nelson, Bob Richardson, Julie Schecter, David Schorr, Timothy Seldes, Molly Simonds, Lee Smith, and Ed Zahniser.

Some of these book excerpts appeared in *Antaeus, The Atlantic, Harper's, Tikkun, Image,* and *Yale Review,* to whose editors I am grateful.

THE ABUNDANCE

TOTAL ECLIPSE

I

IT HAD BEEN LIKE DYING, that sliding down the mountain pass. It had been like the death of someone, irrational, that sliding down the mountain pass and into the region of dread. It was like slipping into fever, or falling down that hole in sleep from which you wake yourself whimpering. We had crossed the mountains that day, and now we were in a strange place—a hotel in central Washington, in a town near Yakima. The eclipse we traveled here to see would occur early next morning.

I lay in bed. My husband, Gary, was reading beside me. I lay in bed and looked at the print on the hotel room wall. It was a print of a detailed and lifelike painting, clearly contemporary, of a smiling clown's head made all out of vegetables—the old Arcimboldo idea, but junk. It was an image of the sort which you do not intend to look at, and which, alas, you never forget. Some tasteless fate presses it upon you; it becomes part of the complex interior junk you carry with you wherever you go. Two years have passed since the total eclipse of which I write. During those years

I have forgotten, I assume, a great many things I wanted to remember—but I have not forgotten that clown print or its lunatic setting in the old hotel.

The bright clown was bald. Actually, he wore a clown's tight rubber wig, painted white; this stretched over the top of his skull, which was a cabbage. His hair was bunches of baby carrots. Inset in his white clown makeup, and in his cabbage skull, were his small and laughing human eyes. The clown's glance was like the glance of Rembrandt in the late self-portraits: lively, knowing, deep, and loving. The crinkled shadows around his eyes were string beans. His eyebrows were parsley. Each of his ears was a broad bean. His thin, joyful lips were red chili peppers; between his lips were wet rows of human teeth and a suggestion of real tongue. The clown print was framed in gilt and glassed.

To put ourselves in the path of the total eclipse, that day we had driven five hours inland from the Washington coast where we lived. When we tried to cross the Cascades range, an avalanche blocked the pass. A slope's worth of snow blocked the road; traffic backed up. Had the avalanche buried any cars that morning? We could not learn. This highway was the only winter road over the mountains.

We waited as highway crews bulldozed a passage through the avalanche. With two-by-fours and walls of plyboard, they erected a one-way, roofed tunnel through the avalanche. We drove through the avalanche tunnel, crossed the pass, and descended several thousand feet into central Washington and the broad Yakima valley, about which we knew only that it was orchard country. As we lost altitude, the snows disappeared; our ears popped; the trees changed, and in the trees were strange birds. I watched the landscape innocently, like a fool, like a diver in the rapture of the deep who plays on the bottom while his air runs out.

The hotel lobby was a dark, derelict room, narrow as a corridor, airless. We waited on a couch while the manager vanished upstairs to do something unknown to our room. Beside us on an overstuffed chair, absolutely motionless, was a platinum-blond woman in her forties wearing a black silk dress and a strand of pearls. Her long legs were crossed; she supported her head on her fist. At the dim far end of the room, their backs toward us, sat six bald old men in their shirtsleeves, around a loud television. Two of them seemed asleep. They were drunks. "Number six!" cried the man on television. "Number six!"

On the broad lobby desk, lighted and bubbling, was a ten-gallon aquarium containing one large fish; the fish tilted up and down in its water. Against the long opposite wall sang a live canary in its cage. Beneath the cage, among spilled millet seeds on the carpet, were a decorated child's sand bucket and matching sand shovel.

The alarm was set for six. I lay awake remembering an article I had read downstairs in the lobby, in an engineering magazine. The article was about gold mining. In South Africa, in India, and in South Dakota, the gold mines extend so deeply into the earth's crust that they are hot. The rock walls burn the miners' hands. The companies have to air-condition the mines; if the air conditioners break, the miners die. The elevators in the mine shafts run very slowly, down, and up, so the miners' ears will not pop in their skulls. When the miners return to the surface, their faces are deathly pale.

Early the next morning we checked out. It was February 26, 1979, a Monday morning. We would drive out of town, find a hilltop, watch the eclipse, and then drive back over the mountains and home to the coast. How familiar things are here; how adept we are; how smoothly and professionally we check out! Gary put the car in gear and off we went, as off we have gone to a hundred other adventures.

II

IT WAS BEFORE DAWN when we found a highway out of town and drove into the unfamiliar countryside. By the growing light we could see a band of cirrostratus clouds in the sky. Later the rising sun would clear these clouds before the eclipse began. We drove at random until we came to a range of unfenced hills. We pulled off the highway, bundled up, and climbed one of these hills.

The hill was five hundred feet high. Long winter-killed grass covered it, as high as our knees. We climbed and rested, sweating in the cold; we passed clumps of bundled people on the hillside who were setting up telescopes and fiddling with cameras. The top of the hill stuck up in the middle of the sky. We tightened our scarves and looked around.

East of us rose another hill like ours. Between the hills, far below, was the highway that threaded south into the valley. This was the Yakima valley; I had never seen it before. It is justly famous for its beauty, like every planted valley. It extended south into the horizon, a distant dream of a valley, a Shangri-la. All its hundreds of low, golden slopes bore orchards. Among the orchards were towns and roads and plowed and fallow fields. Through the valley wandered a thin, shining river, and from the river extended fine, frozen irrigation ditches. Distance blurred and blued the sight, so that the whole valley looked like a thickness or sediment at the bottom of the sky. Directly behind us was more sky, and empty lowlands blued in the distance, and Mount Adams—an enormous, snow-covered volcanic cone rising flat, like so much stage scenery.

Now the sun was up. We could not see it; but the sky behind the band of clouds was yellow, and far down the valley, some hillside orchards had lighted up. More people were parking near the highway and climbing the hills. It was the West. All of us rugged individualists were wearing knit caps and blue nylon parkas. People were climbing the nearby hills and setting up shop in clumps among the dead grasses. It looked as though we had all gathered on hilltops to pray for the world on its last day. It looked as though we had all crawled out of spaceships and were preparing to assault the valley below. It looked as though we were scattered on hilltops at dawn to sacrifice virgins, make rain, set stone stelae in a ring. There was no place out of the wind. The straw grasses banged our legs.

Up in the sky where we stood the air was lusterless yellow. To the west the sky was blue. Now the sun cleared the clouds. We cast rough shadows on the blowing grass; freezing, we waved our arms. Near the sun, the sky was bright and colorless. There was nothing to see.

It began with no ado. Odd that such a well-advertised public event should have no starting gun, no overture, no introductory speaker. I should have known right then that I was out of my depth. Without pause or preamble, silent as orbits, a piece of the sun went away. We looked at it through welders' goggles. A piece of the sun was missing; in its place we saw empty sky.

I had seen a partial eclipse in 1970. A partial eclipse is very interesting. It bears almost no relation to a total eclipse. Seeing a partial eclipse bears the same relation to seeing a total eclipse as kissing a man does to marrying him, or as flying in an airplane does to falling out of an airplane. Although the one experience precedes the other, it in no way prepares you for it. During a partial eclipse the sky does not darken—not even when 94 percent of the sun is hidden. Nor does the sun, seen colorless through protective devices, seem terribly strange. We have all seen a sliver of light in the sky, since we have all seen the crescent moon by day. During a partial eclipse the air does get cold, precisely as if someone were standing between you and the fire. And blackbirds do fly back to their roosts. It is strange enough. I had seen a partial eclipse before, and here it seemed was another.

What you see in a total eclipse is entirely different from what you know. It is especially different for those of us whose grasp of astronomy is so frail that, given a flashlight, a grapefruit, two oranges, and fifteen years, we still could not figure out which way to set the clocks for daylight saving time. Usually it is a bit of a trick to keep your knowledge from blinding you. But during an eclipse it is easy. What you see is much more convincing than any wild-eyed theory you may know.

You may read that the moon has something to do with eclipses. I have never seen the moon yet. You do not see the moon. So near the sun, it is as completely invisible as the stars are by day. What you see before your eyes is the sun going through phases. It gets narrower and narrower, as the waning moon does, and, like the ordinary moon, it travels alone in the simple sky. The sky is of course background. It does not appear to eat the sun; it is far behind the sun. The sun simply shaves away; gradually, you see less sun and more sky.

The sky's blue was deepening, but there was no darkness. The sun was a wide crescent, like a segment of tangerine. The wind freshened and blew steadily over the hill. The eastern hill across the highway grew dusky and sharp. The towns and orchards in the valley to the south were dissolving into the blue light. Only the thin band of river held a spot of sun.

Now the sky to the west deepened to indigo, a color never seen. A dark sky usually loses color. This was saturated, deep indigo, up in the air. Stuck up into that unworldly sky was the cone of Mount Adams, and the alpenglow was upon it—that red light of sunset which holds out on snowy mountaintops long after the valleys and tablelands have dimmed. "Look at Mount Adams," I said, and that was the last sane moment I remember.

I turned back to the sun. It was going. The sun was going, and the world was wrong. The grasses were wrong; they were now platinum. Their every detail of stem, head, and blade shone lightless and artificially distinct as an art photographer's platinum print. This color has never been seen on earth. The hues were metallic; their finish was matte. The hillside was a nineteenth-century tinted photograph from which the tints had faded. All the people you see in the photograph, distinct and detailed as their faces look, are now dead. The sky was navy blue. My hands were silver. All the distant hills' grasses were fine-spun metal which the wind laid down. I was watching a faded color print of a movie filmed in the Middle Ages; I was standing in it, by some mistake. I was standing in a movie of hillside grasses filmed in the Middle Ages. I missed my own century, the people I knew, and the real light of day.

I looked at Gary. He was in the film. Everything was lost. He was a platinum print, a dead artist's version of life. I saw on his skull the darkness of night mixed with the colors of day. My mind was going out; my eyes were receding the way galaxies recede to the rim of space. Gary was light-years away, gesturing inside a circle of darkness, down the wrong end of the telescope. He smiled as if he saw me; the stringy crinkles around his eyes moved. The sight of him, familiar and wrong, was something I was remembering from centuries hence, from the other side of death: Yes, *that* is the way he used to look, when we were living. When it was our generation's turn to be alive. I could not hear him; the wind was too loud. Behind him the sun was going. We had all started down a chute of time. At first it was pleasant; now there was no stopping it. Gary was chuting away across space, moving and talking and catching my eye, chuting down the long corridor of separation. The skin on his face moved like thin bronze plating that would peel.

The grass at our feet was wild barley. It was the wild

einkorn wheat which grew on the hilly flanks of the Zagros Mountains, above the Euphrates valley, above the valley of the river we called *River.* We harvested the grass with stone sickles, I remember. We found the grasses on the hillsides; we built our shelter beside them and cut them down. That is how he used to look then, with the sky so dark behind him, and the wind blowing. God save our life.

From all the hills came screams. A piece of sky beside the crescent sun was detaching, a loosened circle of evening sky, suddenly lighted from the back. It was an abrupt black body out of nowhere; it was a flat disk; it was almost over the sun. That's when the screams began. All at once this disk of sky slid over the sun like a lid. The sky snapped over the sun like a lens cover. The hatch in the brain slammed.

Abruptly it was dark night, on the land and in the sky. In the night sky was a tiny ring of light. For the hole where the sun belongs is very small. Just a thin ring of light marked its place. There was no sound. The eyes dried, the arteries drained, the lungs hushed. There was no world. We were the world's dead people rotating and orbiting around and around, embedded in the planet's crust, while the earth tolled down. Our minds were light-years distant, forgetful of almost everything. Only an extraordinary act of will could recall to us our former, living selves and our contexts in matter and time. We had, it seems, loved the planet and loved our lives, but could no longer remember the way of them. The light was wrong. In the sky was something that should not be there. In the black sky was a ring of light. It was a thin ring, an old, thin silver wedding band, an old, worn ring. It was an old wedding band in the sky, or a morsel of bone. There were stars. It was over.

III

IT IS NOW that the temptation is strongest to leave these regions. We have seen enough; let's go. Why burn our hands any more than we have to? But two years have passed; the price of gold has risen. I return to the same buried alluvial beds and pick through the strata again.

I saw, early in the morning, the sun diminish against a backdrop of sky. I saw a circular piece of that sky appear, suddenly detached, blackened, and backlighted; from nowhere it came and overlapped the sun. It did not look like the moon. It was enormous and black. If I had not read that it was the moon, I could have seen the sight a hundred times and never once thought of the moon. (If, however, I had not read that it was the moon—if, like most of the world's people throughout time, I had simply glanced up and seen this thing—then doubtless I would not have speculated much but, like Emperor Louis of Bavaria in 840, simply died of fright on the spot.) It did not look like a dragon, although it looked more like a dragon than the moon. It looked like a lens cover, or the lid of a pot. It materialized out of thin air—black, and flat, and sliding, outlined in flame.

Seeing this black body was like seeing a mushroom cloud. The meaning of the sight overwhelmed its fascination. It obliterated meaning itself. If you were to glance out one day and see a row of mushroom clouds rising on the horizon, you would know at once that what you were seeing, remarkable as it was, was intrinsically not worth remarking. No use running to tell anyone. Significant as this dread sight was, it would not matter a whit. For what is significance? It is significance for people. No people; no significance. This is all I have to tell you.

In the deeps are the violence and terror of which psychology has warned us. But if you ride these monsters deeper down, if you drop with them farther over the world's rim, you find what our sciences cannot locate or name, the substrate, the ocean or matrix or ether that buoys the rest, that gives goodness its power for good, and evil its power for evil, the unified field: our complex and inexplicable caring for each other, and for our life together here. This is given. It is not learned.

The world that lay under darkness and stillness following the closing of the lid was not the world we know. The event was over. Its devastation lay round about us. The clamoring mind and heart stilled, almost indifferent, certainly disembodied, frail, and exhausted. The hills were hushed, obliterated. Up in the sky, like a crater from some distant cataclysm, was a hollow ring.

You have seen photographs of the sun taken during a total eclipse. The corona fills the print. All of those photographs were taken through telescopes. The lenses of telescopes and cameras can no more cover the breadth and scale of the visual array than language can cover the breadth and simultaneity of internal experience. Lenses enlarge the sight, omit its context, and make of it a pretty and sensible picture, like something on a Christmas card. I assure you, if you send any shepherds a Christmas card on which is printed a three-by-five photograph of the angel of the Lord, the glory of the Lord, and the multitude of the heavenly host, they will not be sore afraid. More fearsome things can come in envelopes. More moving photographs than those of the sun's corona can appear in magazines. But I pray you will never see anything more awful in the sky.

You see the wide world swaddled in darkness; you see a vast breadth of hilly land, and an enormous, distant, blacked valley; you see towns' lights, a river's path, and blurred portions of your hat and scarf; you see your husband's face looking like an early black-and-white film; and you see a sprawl of black sky and blue sky together, with unfamiliar stars in it, some barely visible bands of cloud, and over there, a small white ring. The ring is as small as one goose in a flock of migrating geese—if you happen to notice a flock of migrating geese. It is one 360th part of the visible sky. The sun we see is less than half the diameter of a dime held at arm's length.

The Crab Nebula, in the constellation Taurus, looks, through binoculars, like a smoke ring. It is a star in the process of exploding. Light from its explosion first reached the earth in 1054; it was a supernova then, and so bright it shone in the daytime. Now it is not so bright, but it is still exploding. It expands at the rate of seventy million miles a day. Yet it does not budge. Its apparent size does not increase. Photographs of the Crab Nebula taken fifteen years ago seem identical to photographs of it taken yesterday. Some lichens are similar. Botanists have measured some ordinary lichens twice, at fifty-year intervals, without detecting any growth at all. And yet their cells divide; they live.

The small ring of light was like these things—like a ridiculous lichen up in the sky, like a perfectly still explosion 4,200 light-years away: It was interesting, and lovely, and in witless motion, and it had nothing to do with anything.

It had nothing to do with anything. The sun was too small, and too cold, and too far away, to keep the world alive. The white ring was not enough. It was feeble and worthless. It was as useless as a memory; it was as off kilter and hollow and wretched as a memory.

When you try your hardest to recall someone's face, or the look of a place, you see in your mind's eye some vague and terrible sight such as this. It is dark; it is insubstantial; it is all wrong.

The white ring and the saturated darkness made the earth and sky look as they must look in the memories of the careless dead. What I saw, what I seemed to be standing in, was all the wrecked light that the memories of the dead could shed upon the living world. We had all died in our boots on the hilltops of Yakima, and were alone in eternity. Empty space stoppered our eyes and mouths; we cared for nothing. We remembered our living days wrong. With great effort we recalled some sort of circular light in the sky, but only the outline. And then the orchard trees withered, the ground froze, the glaciers slid down the valleys and overcame the towns. If there had ever been people on earth, nobody knew it. The dead had forgotten those they loved. Parted one from the other, they could no longer remember the faces and lands they had loved in the light. They just stood on the darkened hilltops, looking down.

IV

WE TEACH OUR CHILDREN one thing only, as we were taught: to wake up. We teach our children to look alive there, to join by words and activities the life of human culture on the planet's crust. As adults we are almost all adept at waking up. We have so mastered the transition we forgot we ever learned it. Yet it is a transition we make a hundred times a day, as, like so many willless dolphins, we plunge and surface, lapse and emerge. We live half our waking lives and all of our sleeping lives in some private, useless, and insensible waters we never mention or recall. Useless, I say. Valueless, I might add—until someone hauls their wealth up to the surface and into the wide-awake city, in a form that people can use.

I do not know how we got to the restaurant. Like Roethke, "I take my waking slow." Gradually I seemed more or less alive, and already forgetful. It was now almost nine in the morning. It was the day of a solar eclipse in central Washington, and a fine adventure for everyone. The sky was clear; there was a fresh breeze out of the north.

The restaurant was a roadside place with tables and booths. The other eclipse-watchers were there. From our booth we could see their cars' California license plates, their University of Washington parking stickers. Inside the restaurant we were all eating eggs or waffles; people were fairly shouting and exchanging enthusiasms, like fans after a World Series game. Did you see . . . ? Did you see . . . ? Then somebody said something that knocked me for a loop.

A college student, a boy in a blue parka who carried a Hasselblad, said to us, "Did you see that little white ring? It looked like a Life Saver. It looked like a Life Saver up in the sky."

And so it did. The boy spoke well. He was a walking alarm clock. I myself had at that time no access to such a word. He could write a sentence, and I could not. I grabbed that Life Saver and rode it to the surface. And I had to laugh. I had been dumbstruck on the Euphrates River, I had been dead and gone and grieving, all over the sight of something which, if you could claw your way up to that level, you would grant looked very much like a Life Saver. It was good to be back among people so clever; it was good to have all the world's words at the mind's disposal, so the mind could begin its task. All those things for which we have no words are lost. The mind—the culture—has two little tools, grammar and lexicon: a decorated sand bucket and a matching shovel. With these we bluster about the continents and do all the world's work. With these we try to save our very lives.

There are a few more things to tell from this level, the level of the restaurant. One is the old joke about breakfast. "It can never be satisfied, the mind, never." Wallace Stevens wrote that, and in the long run he was right. The mind wants to live forever, or to learn a very good reason why not. The mind wants the world to return its love, or its awareness; the mind wants to know all the world, and all eternity, even God. The mind's sidekick, however, will settle for two eggs over easy. The dear, stupid body is as easily satisfied as a spaniel. And, incredibly, the simple spaniel can lure the brawling mind to its dish. It is everlastingly funny that the proud, metaphysically ambitious, clamoring mind will hush if you give it an egg.

Further: While the mind reels in deep space, while the mind grieves or fears or exults, the workaday senses—in ignorance or idiocy, like so many computer terminals printing our market prices while the world blows up—still transcribe their little data and transmit them to the warehouse in the skull. Later, under the tranquilizing influence of fried eggs, the mind can sort through all of these data.

The restaurant was a halfway house, a decompression chamber. There I remembered a few things more. The deepest, and most terrifying, was this: I have said that I heard screams. (I have since read that screaming, with hysteria, is a common reaction even to expected total eclipses.) People on all the hillsides, including, I think, myself, screamed when the black body of the moon detached from the sky and rolled over the sun. But something else was happening at that same instant, and it was this, I believe, that made us scream.

The second before the sun went out we saw a wall of dark shadow come speeding at us. We no sooner saw it than it was upon us, like thunder. It roared up the valley. It slammed our hill and knocked us out. It was the monstrous swift shadow cone of the moon. I have since read that this wave of shadow moves 1,800 miles an hour.

Language can give no sense of this sort of speed. It was 195 miles wide. No end was in sight. You saw only the edge. It rolled at you across the land at 1,800 miles an hour, hauling darkness behind it like plague. Seeing it, and knowing it was coming straight for you, was like feeling a slug of anesthetic shoot up your arm. If you think very fast, you may have time to think, "Soon it will hit my brain." You can feel the appalling, inhuman speed of your own blood. We saw the wall of shadow coming, and screamed as it hit.

This was the universe about which we had read so much and never before felt: the universe as a clockwork of loose spheres flung at stupefying, unauthorized speeds. How could anything moving so fast not crash, not veer from its orbit amok like a car out of control on a turn?

Less than two minutes later, when the sun emerged, the trailing edge of the shadow cone sped away. It coursed down our hill and raced eastward over the plain, faster than the eye could believe; it swept over the plain and dropped over the planet's rim in a twinkling. It had clobbered us, and now it roared away. We blinked in the light. It was as though an enormous, loping god in the sky had reached down and slapped the earth's face.

Something else, something more ordinary, came back to me along about the third cup of coffee. During the moments of totality, it was so dark that drivers on the highway below turned on their cars' headlights. We could see the highway's route as a strand of lights. It was bumper-to-bumper down there. It was eight-fifteen in the morning, Monday morning, and people were driving into Yakima to work. That this was as dark as night, and eerie as hell, an hour after dawn, apparently meant that in order to *see* to drive to work, people had to use their headlights. Four or five cars pulled off the road. The rest, though, in a line at least five miles long, drove on into town. The highway ran between hills; the people could not have seen any of the eclipsed sun at all. Yakima will have another total eclipse in 2039. Perhaps, in 2039, businesses will give their employees an hour off.

From the restaurant we drove back to the coast. The highway crossing the Cascades range was open. We drove over the mountain like old pros. We joined our places on the planet's thin crust; it held. For the time being, we were home free.

Early that morning at six, when we had checked out, the six bald men were sitting on folding chairs in the dim hotel lobby. The television was on. Most of them were awake. You might drown in your own spittle, God knows, at any time; you might wake up dead in a small hotel, a cabbage head watching TV while snow piles up in the passes, watching TV while the chili peppers smile and the moon passes over the sun and nothing changes and nothing is learned because you have lost your bucket and shovel and no longer care. What if you regain the surface and open your sack and find, instead of treasure, a beast which jumps at you? Or you may not come back at all. The winches may jam, the scaffolding buckle, the air-conditioning collapse. You may glance up one day and see by your headlamp the canary for pearls and touch a moray eel. You yank on your rope; it's too late.

Apparently people share a sense of these hazards, for when the total eclipse ended, an odd thing happened. When the sun appeared as a blinding bead on the ring's side, the eclipse was over. The black lens cover appeared again, backlighted, and slid away. At once the yellow light made the sky blue again; the black lid dissolved and vanished. The real world began there. I remember now: We all hurried away. We were born and bored at a stroke. We rushed down the hill; we found our car; we saw the other people streaming down the hillsides; we joined the highway traffic and drove away.

We never looked back. It was a general vamoose,

and an odd one, for when we left the hill, the sun was still partially eclipsed—a sight rare enough in itself that we probably would have driven five hours to see it. But enough is enough. One turns at last even from glory itself with a sigh of relief. From the depths of mystery, and even from the heights of splendor, we bounce back and hurry for the latitudes of home.

THE DEER AT PROVIDENCIA

THERE WERE FOUR OF US North Americans in the jungle, in the Ecuadorian jungle on the banks of the Napo River in the Amazon watershed. The other three North Americans were big-city men. We stayed in tents in one riverside village, and visited others. At the village called Providencia we saw a sight that moved all of us, but shocked the men.

The first thing we saw when we climbed the riverbank to the village of Providencia was the deer. It was roped to a tree on the grass clearing near the thatch shelter where we would eat. The deer was small, about the size of a whitetail fawn, and apparently full-grown. It had a rope around its neck and three feet caught in the rope. Someone said that the dogs had caught it that morning and the villagers were going to cook and eat it that night.

This clearing lay at the edge of the little thatched-hut village. We could see the villagers going about their business, scattering feed corn for hens near their houses, and wandering down paths to the river to bathe. The village headman was our host; he stood beside us as we watched the deer struggle. Several village boys were interested in the deer; they formed part of the circle we made around it in the clearing. So also did the four men from Quito who were guiding us around the jungle. Pepe was the real guide. Few of the very different people standing in this circle had a common language. We watched the deer, and no one said much.

The deer lay on its side at the rope's very end, so the rope lacked slack to let it rest its head in the dust. It was "pretty," delicate of bone like all deer, and thin-skinned for the tropics. Its skin looked virtually hairless, in fact, and almost translucent, like a membrane. Its neck was no thicker than my wrist; it had been rubbed open on the rope, and gashed. Trying to get itself free of the rope, the deer had cut its own neck with its hooves. The raw underside of its neck showed red stripes and some bruises bleeding inside the skin. Now three of its feet were hooked in the rope under its jaw. It could not stand, of course, on one leg, so it could not move to slacken the rope and ease the pull on its throat and enable it to rest its head.

Repeatedly the deer paused, motionless, its eyes veiled, with only its rib cage in motion, and its breaths the only sound. Then, after I would think, "It has given up; now it will die," it would heave. The rope twanged; the tree leaves clattered; the deer's free foot beat the ground again. We stepped back and held our breaths. It thrashed, kicking, but with only the one leg. The other three legs tightened inside the rope's loop. Its hip jerked; its spine shook. Its eyes rolled; its tongue, thick with spit, pushed in and out. Then it rested again. We watched this for fifteen minutes.

At one point three young village boys charged in to release its trapped legs, then jumped back to the circle of people. But instantly the deer scratched up its neck with its hooves again and snared its forelegs right back in the rope. It was like Br'er Rabbit and the Tar Baby.

We watched the deer from the circle, and then we drifted on to lunch. Our palm-roofed shelter stood on a grassy promontory from which we could see the deer tied to the tree, pigs and hens walking under village houses, and black-and-white cattle standing in the river. There was a slight breeze.

Lunch, which was the second and better lunch we had that day, was hot and fried. There was a big fish called *doncella,* a kind of catfish, dipped whole in corn flour and beaten egg, then deep-fried. With our fingers we pulled soft fragments of it from its sides to our plates, and ate; it was delicate fish-flesh, fresh and mild. Someone found the roe, and I ate of that too—it was fat and stronger, like egg yolk, naturally enough, and warm.

There was also a stew of meat in shreds with rice and pale brown gravy. I had asked what kind of deer it was tied to the tree; Pepe had answered in Spanish, "*Gama.*" Now they told us this, too, was *gama,* stewed. I suspect the word means merely game. At any rate, I heard that the village dogs had cornered another deer just yesterday, and it was this deer which we were now eating in full sight of the whole article. It was good. I was surprised at its tenderness. But it is a fact that high levels of lactic acid, which builds up in muscle tissues during exertion, tenderizes.

After the fish and meat we ate bananas fried in chunks and served on a tray; they were sweet and full of flavor. I felt terrific. My shirt was wet and cool from swimming; I had had a night's sleep, two decent walks, three meals, and a swim—everything tasted good.

From time to time, one or another of us would look beyond our shaded roof to the sunny spot where the deer was still convulsing in the dust. Finally, our meal completed, we walked around the deer and back to the boats.

That night I learned that while all of us had been watching the deer, the others were also watching me.

We four North Americans had grown close in the jungle in a way that was not the usual artificial intimacy of travelers. We liked one another. We stayed up all that night talking, murmuring, as though we rocked on hammocks slung above time. The others—from big cities: New York, Washington, Boston—remarked now on the lack of expression on my face earlier, as I watched the deer, or the lack, at any rate, of any expression they expected. They had looked to see how I, the youngest of us and the only woman, was taking the sight of the deer's struggles. I looked detached, I don't know. I was thinking. I remember feeling very old and energetic.

I might have said that, like Thoreau, I have traveled widely in Roanoke, Virginia. I eat meat. These things are not issues. They are mysteries.

Gentlemen of the city, what surprises you? That there is suffering here, or that I know it?

We lay in the tent and talked. "If it had been my wife," one man said with special vigor, amazed, "she wouldn't have cared *what* was going on; she would have dropped *everything* right at that moment and gone in the village from here to there to there, she would not have *stopped* until that animal was out of its suffering one way or another. She couldn't *bear* to see a creature in agony like that."

I nodded.

Now I am home. When I wake I comb my hair before the mirror above my dresser. Every morning for the past two years I see in that mirror, beside my sleep-softened face, the blackened face of a burnt man. It is a wire-service photograph clipped from a newspaper and taped to my mirror. The caption reads: "Alan McDonald in Miami hospital bed." All you can see in the photograph is a smudged triangle of face from his eyelids to his lower lip; the rest is bandages. You cannot see the expression in his eyes; the bandages shade them.

The story, headed "Man Burned for Second Time," begins:

> "Why does God hate me?" Alan McDonald asked from his hospital bed.
>
> "When the gunpowder went off, I couldn't believe it," he said. "I just couldn't believe it. I said, 'No, God couldn't do this to me again.' "

He was in a burn ward in Miami, in serious condition. I do not even know if he lived. I wrote him a letter at the time, cringing.

He had been burned before, thirteen years previously, by flaming gasoline. For years he had been having his body restored and his face remade in dozens of operations. He had been a boy, and then a burnt boy. He had already been stunned by what could happen, by how life could veer.

Once I read that people who survive bad burns tend to go crazy; they have a high suicide rate. Medicine cannot ease their pain; drugs just leak away, soaking the sheets, because there is no skin to hold them in. The people just lie there and weep. Later they kill themselves. They had not known, before they were burned, that the world included such suffering, that life could permit them such pain.

This time a bowl of gunpowder had exploded on McDonald. "I didn't realize what had happened at first," he recounted. "And then I heard that sound from thirteen years ago. I was burning. I rolled to put the fire out and I thought, 'Oh God, not again.'

"If my friend hadn't been there, I would have jumped into a canal with a rock around my neck."

His wife concludes the piece, "Man, it just isn't fair."

I read the whole clipping again every morning. This is the Big Time here, every minute of it. Will someone please explain to Alan McDonald in his dignity, to the deer at Providencia in his dignity, what is going on? And copy me on it.

When we walked by the deer at Providencia for the last time, I said to Pepe, with a glance at the deer, "*Pobrecito*"—"Poor little thing." But I was trying out Spanish. I knew at the time it was a ridiculous thing to say.

THE WEASEL

A WEASEL IS WILD. Who knows what he thinks? He sleeps in his underground den, his tail draped over his nose. Sometimes he lives in his den for two days without leaving. Outside, he stalks rabbits, mice, muskrats, and birds, killing more bodies than he can eat warm, and often dragging the carcasses home. Obedient to instinct, he bites his prey at the neck, either splitting the jugular vein at the throat or crunching the brain at the base of the skull, and he does not let go. One naturalist refused to kill a weasel who was socketed into his hand deeply as a rattlesnake. The man could in no way pry the tiny weasel off, and he had to walk half a mile to water, the weasel dangling from his palm, and soak him off like a stubborn label.

And once, says Ernest Thompson Seton—once, a man shot an eagle out of the sky. Examining the eagle, he found the dry skull of a weasel fixed by the jaws to the bird's throat. The supposition is that the eagle had pounced on the weasel and the weasel swiveled and bit as instinct taught him, tooth to neck, and nearly won. I would like to have seen that eagle from the air a few weeks or months before he was shot: Was the whole weasel still attached to his feathered throat, a fur pendant? Or did the eagle eat what he could reach, gutting the living weasel with his talons before his breast, bending his neck to clean the airborne bones?

I have been thinking about weasels because I saw one last week. I startled a weasel who startled me, and we exchanged a long glance.

Near my house in Virginia is a pond—Hollins Pond. It covers two acres of bottomland near Tinker Creek with six inches of water and six thousand lily pads. There is a fifty-five-mile-per-hour highway at one end of the pond, and a nesting pair of wood ducks at the other. Under every bush is a muskrat hole or a beer can. The far end is an alternating series of fields and woods—fields and woods, threaded everywhere with motorcycle tracks—in whose bare clay wild turtles lay eggs.

One evening last week at sunset, I walked to the pond and sat on a downed log near the shore. I was watching the lily pads at my feet tremble and part over the thrusting path of a carp. A yellow warbler appeared to my right and flew behind me. It caught my eye; I swiveled around—and the next instant, inexplicably, I was looking down at a weasel, who was looking up at me.

Weasel! I had never seen one wild before. He was ten inches long, thin as a curve, a muscled ribbon, brown as fruitwood, soft-furred, alert. His face was fierce, small and pointed as a lizard's; he would have made a good arrowhead. There was just a dot of chin, maybe two brown hairs' worth, and then the pure white fur spreading down his underside. He had two black eyes I could not see, any more than you see a window.

The weasel was stunned into stillness as he was emerging from beneath an enormous shaggy wild rosebush four feet away. I was stunned into stillness, twisted backward on the tree trunk. Our eyes locked, and someone threw away the key.

Our look was as if two lovers, or deadly enemies, met unexpectedly on an overgrown path when each had been thinking of something else: a clearing blow to the gut. It was a bright blow to the brain, a sudden beating of brains, with all the charge and intimate grate of rubbed balloons. It emptied our lungs. It felled the forest, moved the fields, and drained the pond; the world dismantled and tumbled into that black hole of eyes.

He disappeared. This was just last week, and already

I can't recall what shattered the enchantment. I think I blinked, I think I retrieved my brain from the weasel's brain, and tried to memorize what I was seeing, and the weasel felt the yank of separation, the careening splashdown into real life and urgent current of instinct. He vanished under the wild rose. I waited motionless, my mind suddenly full of data and my spirit with pleadings, but he didn't return.

Please do not tell me about "approach-avoidance conflicts." I tell you I've been in that weasel's brain for sixty seconds, and he was in mine. Brains are private places, muttering through unique and secret tapes—but the weasel and I, for a sweet and shocking time, both plugged into another tape altogether. Can I help it if it was a blank?

What goes on in his brain the rest of the time? What does a weasel think about? He won't say. His journal is tracks in clay, a spray of feathers, mouse blood and bone: uncollected, unconnected, loose-leaf, and blown.

I would like to learn, or remember, how to live. I come to Hollins Pond not so much to learn how to live as, frankly, to forget about it. That is, I don't think I can learn from a wild animal how to live in particular—shall I suck warm blood, hold my tail high, walk with my footprints precisely over the prints of my hands?—but I might learn something of mindlessness, something of the purity of living in the physical senses and the dignity of living without bias or motive. The weasel lives in necessity and we live in choice, hating necessity and dying at the last ignobly in its talons. I would like to live as I should, as the weasel lives as he should: open to time and death painlessly, noticing everything, remembering nothing, choosing the given with a fierce and pointed will.

I missed my chance. I should have gone for the throat. I should have lunged for that streak of white under the weasel's chin and held on, held on through mud and into the wild rose, held on for a dearer life. We could lie under the wild rose wild as weasels, mute and uncomprehending. I could very calmly go wild. I could live two days in the den, curled, leaning on mouse fur, sniffing bird bones, blinking, licking, breathing musk, my hair tangled in the roots of grasses. Down is a good place to go, where the mind is single. Down is out, out of your ever-loving mind and back to your careless senses.

I remember muteness as a prolonged and giddy fast, where every moment is a feast of utterance received. Time and events are merely poured, unremarked, and ingested directly, like blood pulsed into my gut through a jugular vein. Could two live that way? Could two live under the wild rose, and explore by the pond, so that the smooth mind of each is as everywhere present to the other, and as received and as unchallenged, as falling snow?

We could, you know. We can live any way we want. People take vows of poverty, chastity, and obedience—even of silence—by choice. The thing is to stalk your calling in a certain skilled and supple way, to locate the most tender and live spot and plug into that pulse. This is yielding, not fighting. A weasel doesn't "attack" anything; a weasel lives as he's meant to, yielding at every moment to the perfect freedom of single necessity.

I think it would be well, and proper, and obedient, and pure, to grasp your one necessity and not let it go, to dangle from it limp wherever it takes you. Then even death, where you're going no matter how you live, cannot you part. Seize it and let it seize you up aloft even, till your eyes burn out and let your very bones unhinge and scatter, loosened over fields, over fields and woods, lightly, thoughtless, from any height at all, from as high as eagles.

PAGANISM

EVERY DAY IS A GOD, each day is a god, and holiness holds forth in time. I worship each god, I praise each day splintered down, splintered down and wrapped in time like a husk, a husk of many colors spreading at dawn fast over the mountains split.

I wake in a god. I wake in arms holding my quilt, holding me as best they can inside my quilt.

Someone is kissing me—already. I wake, I cry "Oh." I rise from the pillow. Why should I open my eyes?

I open my eyes. The god lifts from the water. His head fills the bay. He is Puget Sound, the Pacific; his breast rises from pastures; his fingers are firs; islands slide wet down his shoulders. Islands slip blue from his shoulders and glide over the water, the empty, lighted water like a stage.

Today's god rises, his long legs flecked in clouds. He flings his arms, spreading colors; he arches, cupping sky in his belly; he vaults, vaulting and spread, holding all and spread on me like skin.

Under the quilt in my knees' crook is a cat. She wakes, too, and turns to bite her metal sutures. The day is real; already, I can feel it click, hear it clicking under my knees.

The day is real; the sky clicks securely in place over the mountains, locks round the islands, snaps slap on the bay. Air fits flush on farm roofs; it rises inside the doors of bars and rubs at dull barn windows. Air clicks up my hand cloven into fingers and wells my ears' holes, whole and entire. I call it simplicity, the way matter is smooth and alone.

I toss the cat. I stand and shake the quilt. "Oh," I cry. "Oh!"

There is a spider, too, in the bathroom, with whom I keep a sort of company. Her little outfit always reminds me of a certain moth I helped to kill. The spider herself is of uncertain lineage, bulbous at the abdomen and drab. Her six-inch mess of a web works, works somehow, works miraculously, to keep her alive and me amazed. The web is in a corner behind the toilet, connecting the tile wall to tile wall and floor, in a place where there is, I would have thought, scant traffic. Yet under the web are sixteen or so corpses she has tossed to the floor.

The corpses appear to be mostly sow bugs, those little armadillo creatures who live to travel flat out in houses, and die round. There is also a new shred of earwig, three old spider skins crinkled and clenched, and two moth bodies, wingless and huge and empty, moth bodies I drop to my knees to see.

Today the earwig shines darkly and gleams, what there is of him: a dorsal curve of thorax and abdomen, and a smooth pair of cerci by which I know his name. Next week, if the other bodies are any indication, he will be shrunken and gray, webbed to the floor with dust. The sow bugs beside him are hollow and lack color, fragile, a breath away from fluff. The spider skins lie on their sides, translucent and ragged: Their legs dry in knots. And the moths are empty moths, stagger against each other, headless, in a mess of arcing strips of chitin like peeling varnish, like a jumble of buttresses for cathedral domes, like nothing resembling moths, so that I should hesitate to call them moths, except that I have seen a moth reduced to a nub.

Two summers ago I was camping alone in the Blue Ridge Mountains in Virginia. I had hauled myself and my gear up there to read, among other things, James Ramsey Ullman's *The Day on Fire,* a novel about Rimbaud that had made me want to write when I was sixteen; I was hoping it would do the same now. So I read, lost, every day sitting under a tree by my tent, while warblers swung in the leaves overhead and bristle worms trailed their inches over the twiggy dirt at my feet; and I read every night by candlelight, while barred owls called in the forest and pale moths massed around my head in the clearing where my light made a ring.

Moths kept flying into the candle. They would hiss and recoil, lost upside down in the shadows among cookware. Or they would singe their wings and fall, and their hot wings, as if melted, would stick to the first thing they touched—a pan, a lid, a spoon—so that the snagged moths could flutter only in tiny arcs, unable to break free. These I could release by a flip with a stick. In the morning I would find my cooking stuff gilded with torn flecks of moth wings, triangles of shiny dust here and there on the aluminum. So I read, and boiled water, and replenished candles, and read on.

One night a moth flew into the candle, was caught, burnt dry, and held. I must have been staring at the candle, or maybe I looked up from the shadow that crossed my page; at any rate, I saw it all. A golden female moth, a biggish one with a two-inch wingspan, flapped into the fire, dropped her abdomen into the wet wax—stuck, flamed, frazzled, and fried in a second. Her moving wings ignited like tissue paper, enlarging the circle of light in the clearing and creating out of the darkness the sudden blue sleeves of my sweater, the green leaves of jewelweed by my side, the ragged red trunk of a pine. At once the light contracted again and the moth's wings vanished in a fine, foul smoke. At the same time her six legs clawed, curled, blackened, and ceased, disappearing utterly. And her head jerked in spasms, making a spattering noise; her antennae crisped and burned away and her heaving mouth parts cracked like pistol fire. When it was all over, her head was, so far as I could determine, gone, gone the long way of her wings and legs. Had she been new, or old? Had she mated and laid her eggs, had she done her work? All that remained was the glowing horn shell of her abdomen and thorax—a fraying, partially collapsed gold tube jammed upright in the candle's round pool.

And then this moth-essence, this spectacular skeleton, began to act as a wick. She kept burning. The wax rose in the moth's body from her soaking abdomen to her thorax to the jagged hole where her head should be, and widened into flame, a saffron-yellow flame that robed her to the ground like any immolating monk. That candle had two wicks, two flames of identical height, side by side. The moth's head was fire. She burned for two hours, until I blew her out.

She burned for two hours without changing, without bending or leaning—only glowing within, like a building fire glimpsed through silhouetted walls, like a hollow saint, like a flame-faced virgin gone to God, while I read by her light, kindled, while Rimbaud in Paris burnt out his brains in a thousand poems, while night pooled wet at my feet.

And that is why I believe those hollow crisps on the bathroom floor are moths. I think I know moths, and fragments of moths, and chips and tatters of utterly empty moths, in any state. How many of you, I asked the people in my long-ago class, how many of you want to give your lives and be writers? I was trembling from coffee, or cigarettes, or the closeness of faces all around me. (Is this what we live for? I thought; is this the only final beauty: the color of any skin in any light, and living, human eyes?) All hands rose to the questions. (You, Nick? Will you? Margaret? Randy? Why do I want them to mean it?) And then I tried to tell them what the choice must mean: You can't be anything else. You must go at your life with a broadax. . . . They had no idea what I was saying. (I have two hands, don't I? And all this energy, for as long as I can remember. I'll do it in the evenings, after skiing, or on the way home from the bank, or after the children are asleep . . .) They thought I was raving again. It's just as well.

I have three candles here on the table which I free

from the plants and light when visitors come. The cat usually avoids them, although once she came too close and her tail caught fire; I rubbed it out before she noticed. The flames move light over everyone's skin, draw light to the faces of my friends. When the people leave I never blow the candles out and after I'm asleep they flame and burn.

The Cascade Range, in these high latitudes, backs almost into the water. There is only a narrow strip before it, an afterthought of foothills and farms sixty miles wide. The mountains wall well. The rest of the country—most of the rest of the planet, in some real sense, excluding a shred of British Columbia's coastline and the Alaskan islands—is called, and profoundly felt to be, simply "East of the Mountains." I've been there.

I came here to study hard things—rock mountain and salt sea—and to temper my spirit on their edges. These mountains—Mount Baker and the Sisters and Shuksan, the Canadian Coastal Range and the Olympics on the Peninsula—are surely the edge of the known and comprehended world. They are high. That they bear their own unimaginable masses and climates aloft, holding them up in the sky for anyone to see plain, makes them only more mysterious for their very visibility and absence of secrecy. They are the western rim of the real, if not considerably beyond it. If the Greeks had looked at Mount Baker all day, would their large and honest art not have cracked? Would they not have gone fishing, as these people do? As perhaps I one day shall.

But the mountains are, incredibly, east. When I first came here I faced east and watched the mountains, thinking, These are the Ultima Thule, the final westering, the last serrate margin of time. And since they are, incredibly, east of me, I must be no place at all. But the sun rose over the snowfields and woke me where I lay, and I rose and cast a shadow over someplace, and thought: There is, God help us, more. So gathering my bowls and spoons, and turning my head, as it were, I moved to face west, relinquishing all hope of sanity, for what is more.

What was more is islands: sea, and unimaginably solid islands, and more sea, and a hundred rolling skies. You spill your breath. Nothing holds; the whole show rolls. I can imagine Virginias no less than Pacifics. Inland valley, pool, desert, plain—it's all a falling sheaf of edges, like a quick-flapped deck of cards, like a dory or a day launched all unchristened, lost at sea. Land is a poured thing and time a surface lapping and fringeing at fastness, at a hundred hollow and receding blues. Breathe fast: We're backing off the rim.

Here is the fringey edge where elements meet and realms mingle, where time and eternity spatter each other with foam. The salt sea and the islands, molding and molding, row upon rolling row, don't quit; nor do winds end nor skies cease from spreading in curves. The actual percentage of land mass to sea in Puget Sound equals that of the rest of the planet. We have less time than we knew. Time is eternity's pale interlinear, as the islands are the sea's. We have less time than we knew and that time buoyant and cloven, lucent, and missile, and wild.

The room where I live is plain as a skull, a firm setting for windows. A nun lives in the fires of the spirit, a thinker lives in the bright wick of the mind, an artist lives jammed in the pool of materials. But this room is a skull, a fire tower, wooden, and empty. Of itself it is nothing, but the view, as they say, is good.

Since I live in one room, one long wall of which is glass, I am myself, at everything I do, a backdrop to all the landscape's occasions, all its weathers, colors, and lights. From the kitchen sink, and from my bed, and from the table, couch, hearth, and desk, I see land and water, islands, sky.

The land is complex and shifting; the eye leaves it.

There is a white Congregationalist church among Douglas firs; there is a green pasture between two yellow fallow fields; there are sheep bent over below some alders, and beside them a yard of brown hens. But everything in the landscape points to sea. The land's progress of colors leads the eye up a far hill, a sweeping big farm of a hill whose pale pastures bounce light all day from a billion stems and blades. Down the hill's rim drops a dark slope of forest, a slant your eye rides down to the point, the dark sliver of land that holds the bay. From this angle you see the bay cut a crescent; your eye flies up the black beach to the point, or slides down the green firs to the point, and the point is an arrow pointing over and over, with its log-strewn beach, its gray singleness, and its recurved white edging of foam, to sea: to bright Haro Strait, the bluing of water with distance at the world's edge, and on it the far blue islands, and over these lights the light clouds.

You can't picture it, can you? Neither can I. Oh, the desk is yellow, the oak table round, the ferns alive, the mirror cold, and I never cared. I read. In the Middle Ages, I read, "the idea of a thing which a man framed for himself was always more real to him than the actual thing itself." Of course. I am in my own Middle Ages; the world at my feet, the world through the window, is an illuminated manuscript whose leaves the wind takes, one by one, whose painted illuminations and halting words draw me, one by one, and I am dazzled in day and lost.

There is, in short, one country, one room, one enormous window, one cat, one spider, and one person: but I am myself hollow. And, for now, there are the many gods of mornings and the many things to give them for their work—lungs and heart, muscle, nerve, and bone—and there is a no-man's-land of many things wherein they dwell, and from which I seek to call them, in work that's mine.

So I read. Armenians, I read, salt their newborn babies. I check somewhere else: so did the Jews at the time of the prophets. They washed a baby in water, salted him, and wrapped him in cloths. When God promised to Aaron and all the Levites all the offerings Israel made to God, the firstfruits and the firstling livestock, "all the best of the oil, and all the best of the wine," he said of this promise, "It is a covenant of salt forever." In the Roman church baptism, the priest once placed salt in the infant's mouth.

I salt my breakfast eggs. All day long I feel created. I can see the blown dust on the skin on the back of my hand, the tiny trapezoids of chipped clay, moistened and breathed alive. There are some created sheep in the pasture below me, sheep set down here precisely, just touching their blue shadows hoof to hoof on the grass. Created gulls pock the air, rip great curved seams in the settled air. I greet my created meal, amazed.

I have been drawing a key to the islands I see from my window. Everyone told me a different set of names for them, until one day a sailor came and named them all with such authority that I believed him. So I penciled an outline of the horizon on a sheet of paper and began labeling the lobes: Skipjack, Sucia, Saturna, Salt Spring, Bare Island . . .

Today, November 18 and no wind; today a veil of air has lifted I didn't know was there. Behind the blue translucence of Salt Spring Island I see a new island, a new wrinkle, the deepening of wonder. I have no way of learning this new island's name. Still, I bring the labeled map to the table and pencil in a new line. Call that: Unknown Island North; Water-Statue; Sky-Ruck; Newborn and Salted; Waiting for Sailor.

Henry Miller relates that Knut Hamsun once said, in response to a questionnaire, that he wrote to kill time.

This is funny in a number of ways. In a number of ways I kill myself laughing, looking out at islands. Startled, the yellow cat on the floor stares over her shoulder. She has carried in a wren, I suddenly see, a wren she has killed, whose dead wings point askew on the circular rug. It is time. Out with you both. I'm busy laughing, to kill time. I shoo the cat from the door, turn the wren over in my palm, and drop it from the porch, down to the winter-killed hair grass and sedge, where the cat may find it if she will, or crows, or beetles, or rain.

When I next look up from my coffee, there is a ruckus on the porch. The cat has dragged in a god, scorched. He is alive. I run outside. Save for his wings, he is a perfect, very small man. He is fair, thin-skinned in the cat's mouth, and kicking. His hair is on fire and stinks; his wingtips are blacked and seared. From the two soft flaps of the cat's tiger muzzle his body jerks, naked. One of his miniature hands pushes hard at her nose. He waves his thigh; he beats her face and the air with his smoking wings. I cannot breathe. I run at the cat to scare her; she drops him, casting me an evil look, and runs from the porch.

The god lies gasping and perfect. He is no longer than my face. Quickly I snuff the smoldering fire in his yellow hair with a finger and thumb. In so doing I accidentally touch his skull, brush again his hot skull, which is the size of a hazelnut, as the saying goes, warm-skinned and alive.

He rolls his colorless eyes toward mine; his long wings catch strength from the sun, and heave.

Later I am walking in the day's last light. The god rides barefoot on my shoulder, astride it, tugging on loops of my hair. He is whistling at my ear; he is blowing a huge tune in my ear, a myth about November. He is heaping a hot hurricane into my ear, into my hair, an ignorant ditty calling things real, calling islands out of the sea, calling solid moss from curling rock, and ducks down the sky for the winter.

I see it! I see it all! Two islands, twelve islands, worlds, gather substance, gather the blue contours of time, and array themselves down distance, mute and hard.

I seem to see a road; I seem to be on a road, walking. I seem to walk on a blacktop road that runs over a hill. The hill creates itself, a powerful suggestion. It creates itself, thickening with apparently solid earth and waving plants, with houses and browsing cattle, unrolling wherever my eyes go, as though my focus were a brush painting in a world. I cannot escape the illusion. The colorful thought persists, this world, a dream forced into my ear and sent round my body on ropes of blood. If I throw my eyes past the rim of the hill to see the real stars, were they? Something with wings? I elaborate the illusion instead; I rough in a middle ground. I stitch the transparent curtain solid with phantom mountains, with thick clouds gliding just so over their shadows on green water, with blank impenetrable sky. The dream fills in, like wind widening over a bay. Quickly I look to the fat dream's rim for a glimpse of that old deep . . . and, just as quickly, the blue slaps shut, the colors wrap everything out. There is not a chink. The sky is gagging on trees. I seem to be on a road, walking, greeting the hedgerows, the rose hips, apples, and thorn. I seem to be on a road walking, familiar with neighbors, high-handed with cattle, smelling the sea, and alone. Already, I know the names of things. I can kick a stone.

Time is enough, more than enough, and matter multiple and given. The god of today is a child, a baby new and filling the house, remarkable here in the flesh. He is day. He thrives in a cup of wind, landlocked and thrashing. He unrolls, revealing his shape an edge at a time, a smatter of content, foot first: a word, a friend for coffee, a windshift, the shingling or coincidence of ideas. Today, November 18 and no wind, is clear. Terry Wean—who fishes, and takes my poetry course—could see Mount Rainier. He hauls his reef net gear from the bay; we talk on its deck while he hammers shrunken knots. The

Moores for dinner. In bed, I call to me my sad cat, and read on. Like a rug or wrap rolling unformed up a loom, the day discovers itself, like the poem.

The god of today is rampant and drenched. His arms spread, bearing moist pastures; his fingers spread, fingering the shore. He is time's live skin; he burgeons up from day like any tree. His legs spread crossing the heavens, flicking hugely, and flashing and arcing around the earth toward night.

This is the one world, bound to itself and exultant. It fizzes up in trees, trees heaving up streams of salt to their leaves. This is the one air, bitten by grackles; time is alone and in and out of mind. The god of today is a boy, pagan and fernfoot. His power is enthusiasm; his innocence is mystery. He sockets into everything that is, and that right holy. Loud as music, filling the grasses and skies, his day spreads rising at home in the hundred senses. He rises, new and surrounding. He *is* everything that is, wholly here and emptied—flung, and flowing, sowing, unseen, and flown.

WAKING UP

CHILDREN TEN YEARS OLD WAKE UP and find themselves here, discover themselves to have been here all along; is this sad? They wake like sleepwalkers, in full stride; they wake like people brought back from cardiac arrest or from drowning: in medias res, surrounded by familiar people and objects, equipped with a hundred skills; they know the neighborhood, they can read and write English; they are old hands at the commonplace mysteries, and yet they feel themselves to have just stepped off the boat, just converged with their bodies, just flown down from a trance, to lodge in an eerily familiar life already well under way.

I woke up in bits, like all children, piecemeal over the years. I discovered myself and the world, and then forgot them, and discovered them again. I woke at intervals, that is, until the intervals of waking tipped the scales, and I was more often awake than not. I noticed this process of waking, and guessed with terrifying logic that one of these years not far away I would be awake continuously and never slip back, and never be free of myself again.

Consciousness converges with the child as a landing tern touches the outspread feet of its shadow on the sand: Precisely, toe hits toe. The tern folds its wings to sit; its shadow dips and spreads over the sand to meet and cup its breast.

Like any child, I slid into myself perfectly fitted, as the diver meets her reflection in a pool. Her fingertips on the water, her wrists slide up her arms. The diver wraps herself in her reflection wholly, sealing it at the toes, and wears it as she climbs rising from the pool, and ever after.

SKIN

OUR PARENTS AND GRANDPARENTS, and all their friends, seemed insensible to their own prominent defect, their limp, coarse skin.

We children had, for instance, proper hands; our fluid, pliant fingers joined their skin. Adults had misshapen, knuckly hands loose in their skin like bones in bags; it was a wonder they could open jars. They were loose in their skins all over, except at the wrists and ankles, like rabbits.

We were whole; we were pleasing to ourselves. Our crystalline eyes shone from firm, smooth sockets; we spoke in pure, piping voices through dark, tidy lips. Adults were coming apart, but they neither noticed nor minded. My revulsion was rude, so I hid it. Besides, we could never rise to the absolute figural splendor they alone could on occasion achieve. Our beauty was a mere absence of decrepitude; their beauty, when they had it, was not passive but earned; it was grandeur; it was a party to power, and to artifice, even, and to knowledge. Our beauty was, in the long run, merely elfin. We could not, finally, discount the fact that in some sense they owned us, as they owned the world.

Mother let me play with one of her hands. She laid it flat on a living-room end table beside her chair. I picked up a transverse pinch of skin over the knuckle of her index finger and let it drop. The pinch didn't snap back; it lay dead across her knuckle in a yellowish ridge. I poked it; it slid over intact. I left it there as an experiment and shifted to another finger. Mother was reading *Time* magazine.

Carefully, lifting it by the tip, I raised her middle finger an inch and released it. It snapped back to the tabletop. Her insides, at least, were alive. I tried all the fingers. They all worked. Some I could lift higher than others.

"That's getting boring."

"Sorry, Mama."

I refashioned the ridge on her index-finger knuckle; I made the ridge as long as I could, using both my hands. Moving quickly, I made parallel ridges on her other fingers—a real mountain chain, the Alleghenies; Indians crept along just below the ridgetops, eyeing the frozen lakes below them through the trees.

Skin was earth; it was soil. I could see, even on my own skin, the joined trapezoids of dust specks God had wetted and stuck with his spit in the morning he made Adam from dirt. Now, all these generations later, we people could still see on our skin the inherited prints of the dust specks of Eden.

I loved this thought, and repeated it for myself often. I don't know where I got it; my parents cited Adam and Eve only in jokes. Someday, with the aid of a mirror, I would count the trapezoids and learn precisely how many dust specks Adam comprised—one single handful God wetted, shaped, and blew into, then set firmly into motion and left to wander about in the fabulous garden, bewildered.

The skin on my mother's face was smooth, fair, and tender; it took impressions readily. She napped on her side on the couch, her face skin pooled on the low side; it piled up in the low corner of her deep-set eyes and drew down her lips and cheeks. How flexible was it? I pushed at a puddle of it by her nose.

She stirred and opened her eyes. I jumped back.

She reminded me not to touch her face while she was sleeping. Anybody's face.

When she sat up, her cheek and brow bone bore a deep red gash, the mark of a cushion's welting. It was textured inside precisely with the upholstery's weave and brocade.

Another day, after a similar nap, I spoke up about this

gash. I told her she had a mark on her face where she'd been sleeping.

"Do I?" she said; she ran her fingers through her hair. Her hair was short, blond, and wavy. She wore it swept back from her high, curved forehead. The skin on her forehead was both tight and soft. It would only barely shift when I tried to move it.

She went to the kitchen. The hideous mark on her face did not interest her.

"It'll go away," I said.

"What?" she called.

I noticed the hair on my father's arms and legs; each hair sprang from a dark dot on his skin. I lifted a hair and studied the puckered tepee of skin it pulled with it. Those hairs were in there tight. The greater the strain I put on the hair, the more puckered the tepee became, and shrunken within, concave. I could point it every which way.

"Ouch! Enough of that."

"Sorry, Daddy."

At the beach I felt my parents' shinbones. The bones were flat and curved, like the slats in a venetian blind.

The long edges were sharp as swords. But they had unexplained and, I thought, possibly diseased irregularities: nicks, bumps, small hard balls, shallow ridges, and soft spots. I was lying between my parents on an enormous towel through which I could feel the hot sand.

Loose under their shinbones hung the relaxed flesh of their calves; you could push these and swing them like a baby in a sling. Their heels were dry and hard, sharp at the curved edge. The bottoms of their toes had flattened, holding the imprint of life's smooth floors even when they were lying down. I would not let this happen to me. Under certain conditions, the long bones of their feet showed under their skin. The bones rose up in skeletal rays on the slopes of their insteps, long and miserable and thin. This terrible sight they ignored also.

In fact, they were young. Mother was twenty-two when I was born, and Father twenty-nine; both appeared to other adults much younger than they were. They were a handsome couple. I felt it overwhelmingly when they dressed for occasions. I never lost a wondering awe at the transformation of an everyday, tender, nap-creased mother into an exalted and dazzling beauty who chatted with me as she dressed.

Her blue eyes shone and caught the light, and so did

the platinum waves in her hair and the pearls at her ears and throat. She was wearing a black dress. The smooth skin on her breastbone rent my heart, it was so familiar and beloved; the black silk bodice and the simple necklace set off its human fineness. Mother was perhaps a bit vain of her long and perfect legs, but not too vain for me; despite her excited pleasure, she did not share my view of her beauty.

"Look at your father," she said. We were all in the dressing room. I found him in one of the long mirrors, where he waggled his outthrust chin over the last push of his tie knot. For me he made his big ears jiggle on his skull. It was a wonder he could ever hear anything; his head was loose inside him.

Father's enormousness was an everyday, stunning fact; he was taller than everyone else. He was neither thin nor stout; his torso was supple, his long legs nimble. Before the dressing-room mirror he produced an anticipatory soft-shoe, and checked to see that his cuffs stayed down.

And then they were off. I hoped they knocked 'em dead; I hoped their friends saw how witty they were, and how splendid.

Their parties at home did not seem very entertaining, although they laughed loudly and often fetched the one-man percussion band from the basement, or an old trumpet, or a snare drum. We children could have shown them how to have a better time. Kick the Can, for instance, never palled. A private game called Spider Cow, played by the Spencer children, also had possibilities: The spider cow hid and flung a wet washcloth at whoever found it, and erupted from hiding and chased him running all over the house.

But implicitly and emphatically, my parents and their friends were not interested. They never ran. They did not choose to run. It went with being old, apparently, and having their skin half off.

BEING CHASED

SOME BOYS TAUGHT ME TO PLAY FOOTBALL. This was fine sport. You thought up a new strategy for every play and whispered it to the others. You went out for a pass, fooling everyone. Best, you got to throw yourself mightily at someone's running legs. Either you brought him down or you hit the ground flat out on your chin, with your arms empty before you. It was all or nothing. If you hesitated in fear, you would miss and get hurt: You would take a hard fall while the kid got away, or, worse, you would get kicked in the face while the kid got away. But if you flung yourself wholeheartedly at the back of his knees—if you gathered and joined body and soul and pointed them diving fearlessly—then you likely wouldn't get hurt, and you'd stop the ball. Your fate, and your team's score, depended on your concentration and courage. Nothing girls did could equal it.

Boys welcomed me at baseball, too, for I had, through enthusiastic practice, what was weirdly known as a boy's arm. In winter, in the snow, there was neither baseball nor football, so the boys and I threw snowballs at passing cars. I got in trouble throwing snowballs, and have seldom been happier since.

One weekday morning after Christmas, six inches of new snow had just fallen. We were standing up to our boot tops in snow on a front yard on well-trafficked Reynolds Street, waiting for cars. The cars traveled Reynolds Street slowly and evenly; they were cream puffs, targets all but wrapped in red ribbons. We couldn't miss.

I was seven; the boys were eight, nine, and ten. The oldest two Fahey boys were there—Mikey and Peter—polite blond boys who lived near me on Lloyd Street, and who already had four brothers and sisters. My parents approved of Mikey and Peter Fahey. Chickie McBride was there, a tough kid, and Billy Paul and Mackie Kean, too, from across Reynolds, where the boys grew up dark and furious, grew up skinny, knowing, and skilled. We had all drifted from our houses that morning looking for action, and had found it here on Reynolds Street.

It was cloudy and cold. The cars' tires laid behind them on the snowy street a complex trail of beige chunks like crenellated castle walls. I had stepped on some earlier; they squeaked. We could have wished for more traffic. When a car came, we all popped it one. In the intervals between cars we reverted to the natural solitude of children.

I started making an iceball—a perfect iceball, from the perfectly white snow, perfectly spherical, and squeezed perfectly translucent, all the way through. (The Fahey boys and I considered it unfair actually to throw an iceball at somebody but it had been known to happen.)

I had just embarked on the iceball project when we heard tire chains come clanking from afar. A black Buick was moving toward us down the street. We all spread out, banged together some regular snowballs, took aim, and, when the Buick drew nigh, fired.

A soft snowball hit the driver's windshield right in front of the driver's face. It made a smashed star with a hump in the middle.

Often, of course, we hit our target, but this time, the only time in all of life, the car pulled over and stopped. Its wide black door opened; a man sprang out of it, running. He didn't even close the car door.

He ran after us, right up the snowy Reynolds sidewalk, and we ran away from him. At the corner, I looked back; incredibly, he was still after us. He was in city clothes: a suit and tie, street shoes. Any normal adult would have quit, having sprung us into flight and made his point. This man, though, was gaining on us. He was a thin man, all action. All of a sudden, we were running for our lives.

Wordless, we split up. We were on our turf; we could lose ourselves in the neighborhood backyards, everyone for himself. I paused and considered. Everyone had vanished except Mikey Fahey, who was just rounding the corner of a yellow brick house. Poor Mikey, I trailed him, and the driver of the Buick sensibly picked the two of us to follow. The man apparently had all day.

He chased Mikey and me around the yellow house and up a backyard path we knew by heart: under a low tree, up a bank, through a hedge, down some snowy steps, and across the grocery store's delivery driveway. We smashed through a gap in another hedge, entered a scruffy backyard, and ran around its back porch and tight between houses to Edgerton Avenue; we ran across Edgerton to an alley and up our own sliding woodpile to the Halls' front yard; he kept coming. We ran up Lloyd Street and wound through mazy backyards toward the steep hilltop at Willard and Lang.

He chased us silently, block after block. He chased us silently over picket fences, through thorny hedges, between houses, around garbage cans, and across streets. Every time I glanced back, choking for breath, I expected he would have quit. He must have been as breathless as we were. His jacket strained over his body. It was an immense discovery, pounding into my hot head with every sliding, joyous step, that this ordinary adult evidently knew what I thought only children who trained at football knew: that you have to fling yourself at what you're doing, you have to point yourself, forget yourself, aim, dive.

Mikey and I had nowhere to go, in our own neighborhood or out of it, but away from this man who was chasing us. He impelled us forward; we compelled him to follow our route. The air was cold; every breath tore my throat. We kept running, block after block; we kept improvising, backyard after backyard, running a frantic course and choosing it simultaneously, failing always to find small places or hard places that might slow him down, and discovering always, exhilarated, dismayed, that only bare speed could save us—for he would never give us up, this man—and we were beginning to lose speed.

Ten blocks he chased us through the backyard labyrinths before he finally caught us by our jackets. He caught us and we all stopped.

We three stood staggering, half-blinded, coughing, in an obscure hilltop backyard, a man in his twenties, a boy, a girl. He had released our jackets, our pursuer, our captor, our hero: for he knew we weren't going anywhere. We all played by the rules. Mikey and I unzipped our jackets. I pulled off my sopping mittens. We looked back over our tracks multiplied in the backyard's new snow. All morning we had been breaking that soft white surface. We didn't look at each other. I was cherishing my excitement. The man's lower pants legs were wet; his cuffs were full of snow, and there was a prow of snow beneath them on his shoes and socks. Some trees bordered the little flat backyard, some messy winter trees. There was no one around: a clearing in a grove, and we the only players.

It was a long time before he could speak. I had some difficulty at first recalling why we were even there. My lips felt swollen; I couldn't see out of the sides of my eyes; I kept coughing.

"You stupid kids," he began perfunctorily.

We listened perfunctorily, too, if we listened at all, for the chewing out was redundant, a mere formality, and beside the point. The point was that he had chased us passionately without giving up, and that he had caught us. I wanted the glory to last forever.

But how could it? We could have run through every backyard in North America until we got to Panama. And when he trapped us at the lip of the Panama Canal? What precisely could he have done to prolong the drama of the chase and cap its glory?

I brooded about this for the next few years. He could have fried Mikey Fahey and me in boiling oil, say, or dismembered us piecemeal, or staked us to anthills.

None of which I really wanted, and none of which any adult was likely to do, even in the spirit of fun. He could only chew us out there in the Panamanian jungle, after months or years of exalting pursuit. He could only begin, "You stupid kids," and continue in his ordinary Pittsburgh accent with his normal righteous anger and the usual common sense.

If in that snowy backyard the driver of the black Buick had cut off our heads, Mikey's and mine, I would have died happy, for nothing since has required so much of me as being chased all over Pittsburgh in the middle of winter—running terrified, exhausted—by this sainted, skinny, furious redheaded man who wished to have a word with us.

I don't know how he found his way back to his car.

JOKES

OUR PARENTS WOULD SOONER HAVE LEFT US out of Christmas than leave us out of a joke. They explained a joke to us while they were still laughing at it; they tore a still-kicking joke apart, so we could see how it worked. When we got the first Tom Lehrer album in 1954, Mother went through the album with me, cut by cut, explaining. BVDs are men's underwear. Radiation makes you sterile, and lead protects from radiation, so the joke is . . .

Our father kept in his breast pocket a little black notebook. There he noted jokes he wanted to remember. Remembering jokes was a moral obligation. People who said, "I can never remember jokes," were like people who said, obliviously, "I can never remember names," or, "I don't bathe."

"No one tells jokes like your father," Mother said. Telling a good joke well—successfully, perfectly—was the highest art. It was an art because it was up to you: If you did not get the laugh, you had told it wrong. Work on it, and do it better next time. Never blame the joke, or, worse, the audience.

As we children got older, our parents discussed with us every technical, theoretical, and moral aspect of the art. We tinkered with a joke's narrative structure: "Maybe you should begin with the Indians." We polished the wording. We analyzed pacing. We admired with Father the leisurely meanders of the shaggy-dog story. "A young couple moved to the Swiss Alps," one story of his began, "with their grand piano"; and ended, to a blizzard of thrown napkins, ". . . Oppernockity tunes but once." "Frog goes into a bank," another story began, to my enduring pleasure. The joke was not great, but with what a sweet light splash you could launch it! "Frog goes into a bank," you said, and your canoe had slipped delicately and surely into the water, into Lake Champlain with painted Indians behind every tree, and there was no turning back.

Father was also very fond of stories set in bars that starred zoo animals or insects. These creatures apparently came into bars all over America, sometimes accompanied, sometimes alone, and sat down to face incredulous, sarcastic bartenders. (It was a wonder the bartenders were always so surprised to see talking dogs or drinking monkeys or performing ants, so surprised year after year, when clearly this sort of thing was the very essence of bar life.) In the years he had been more loose, swinging aloft in the airy interval between college and marriage, Father had frequented bars in New York, listening to jazz. Bars had no place whatever in the small Pittsburgh world he had grown up in, and lived in now. Bars were so far from our experience that I had assumed, in my detective work, that their customers were ipso facto crooks. Father's bar jokes—"and there were the regulars, all sitting around"—gave him the raffish air of a man at home anywhere. (How poignant were his "you knows" directed at me: You know how bartenders are; you know how the regulars would all be sitting around. For either I, a nine-year-old girl, knew what he was talking about, then or ever, or nobody did. Only because I read a lot, I often knew.)

Our mother favored a staccato, stand-up style; if our father could perorate, she could condense. Fellow goes to a psychiatrist. "You're crazy." "I want a second opinion!" "You're ugly." How do you get an elephant out of the theater? "You can't; it's in his blood."

What else in life so required, and so rewarded, such care?

"Tell the girls the one about the four-by-twos, Frank."

"Let's see. Let's see."

"Fellow goes into a lumberyard . . ."

"Yes, but it's tricky. It's a matter of point of view."

And Father left the dining room, rubbing his face as if he were smearing on greasepaint, and returned when he was ready.

"Ready with the four-by-twos?" Mother said.

Our father hung his hands in his pockets and regarded the far ceiling with fond reminiscence.

"Fellow comes into a lumberyard," he began. "Says to the guy, 'I need some four-by-twos.' 'You mean two-by-fours?' 'Just a minute. I'll find out.' He walks out to the parking lot, where his buddies are waiting in the car. They roll down the car window. He confers with them awhile and comes back across the parking lot and says to the lumberyard guy, 'Yes. I mean two-by-fours.'

"Lumberyard guy says, 'How long do you want them?' 'Just a minute,' fellow says, 'I'll find out.' He goes out across the parking lot and confers with the people in the car and comes back across the parking lot to the lumberyard and says to the guy, 'A long time. We're building a house.' "

After any performance Father rubbed the top of his face with both hands, as if it had all been a dream. He sat back down at the dining-room table, laughing and shaking his head. "And when you tell a joke," Mother said to Amy and me, "laugh. It's mean not to."

We were brought up on the classics. Our parents told us all the great old American jokes, practically by number. They collaborated on, and for our benefit specialized in, the painstaking, near-paleontological reconstruction of vanished jokes from extant taglines; they could vivify old *New Yorker* cartoons, source of many taglines. The lines themselves—"Back to the old drawing board," and, "I say it's spinach and I say the hell with it," and, "A simple yes or no will suffice"—were no longer funny; they were instead something better, they were fixtures in the language. The taglines of old jokes were the most powerful expressions we learned at our parents' knees. A few words suggested an entire story and a wealth of feelings. Learning our culture backward this way, Amy and Molly and I heard only later about *The Divine Comedy* and the Sistine Chapel ceiling, and still later about the Greek and Roman myths, which held no residue of feeling for us at all—certainly not the vibrant suggestiveness of old American jokes and cartoons.

Our parents reserved a few select jokes, such as "Archibald à Soulbroke," like vintage wines, for extraordinary occasions. We heard about or witnessed those rare moments—maybe three or four in a lifetime—when circumstances combined to float our father to the top of the world, from which precarious eminence he would consent to fling himself into "Archibald à Soulbroke."

Telling "Archibald à Soulbroke" was for Father an exhilarating ordeal, like walking a tightrope over Niagara Falls. It was a long, absurdly funny, excruciatingly tricky tour de force he had to tell fast, and it required beat-perfect concentration. He had to go off alone and rouse himself to an exalted, superhuman pitch in order to pace the hot coals of its dazzling verbal surface. Often enough he returned to a crowd whose moment had passed.

There was another very complicated sort of joke that required a long weekend with tolerant friends. First, you had to tell a joke that was not funny. It was a long, pointless story about a construction job that ended with someone's throwing away a brick. There was nothing funny about it at all, and when your friends did not laugh, you had to pretend you'd muffed it. (Your husband in the crowd could shill for you: " 'Tain't funny, Pam. You told it all wrong.")

A few days later, if you could contrive another occasion for joke telling, and if your friends still permitted you to speak, you set forth on another joke, this one an old nineteenth-century chestnut about angry passengers on a train: The lady plucks the lighted, smelly cigar from the man's mouth and flings it from the moving train's window. The man seizes the little black poodle from her lap and hurls the poor dog from the same window. When at last the passengers draw unspeaking into the station, what do they see coming down the platform but the black poodle, and guess what it has in its mouth? "The cigar," say your friends, bored sick and vowing to never spend another weekend with you. "No," you say, triumphant, "the brick."

This was Mother's kind of joke. Its very riskiness excited her. It wasn't funny, but it was interesting to set up, and it elicited from her friends a grudging admiration.

How long, I wondered, could you stretch this out? How boldly could you push an audience—not, in Mother's terms, to "slay them," but to please them in some grand way? How could you convince the listeners that you knew what you were doing, that the payoff would come? Or conversely, how long could you lead them to think you were stupid, a dumb blonde, only to enhance their surprise at the punch line, and heighten their pleasure in the good story you had controlled all along? Alone, energetic, and trying to fall asleep, or walking the residential streets long distances every day, I pondered these things.

Our parents were sympathetic to what professional comedians call flop sweat. Boldness was all at our house, and of course you would lose some. Anyone could be misled by poor judgment into telling a "woulda hadda been there." Telling a funny story was harder than telling a joke; it was trying out, as a tidy unit, some raveling shred of the day's fabric. You learned to gauge what sorts of thing would "tell." You learned that some people, notably your parents, could rescue some things by careful narration from the category "woulda hadda been there" to the category "it tells."

At the heart of originating a funny story was recognizing it as it floated by. You scooped the potentially solid tale from the flux of history. Once I overheard my parents arguing over a thirty-year-old story's credit line. "It was my mother who said that," Mother said. "Yes, but"—Father was downright smug—"*I* was the one who noticed she said that."

The sight gag was a noble form, and the running gag was a noble form. In combination they produced the top of the line, the running sight gag. How splendid it was when my parents could get a running sight gag going. We heard about these legendary occasions with a thrill of family pride, as other children hear about their progenitors' war exploits.

The sight gag could blur with the practical joke—not a noble form but a friendly one, that helps the years pass. My parents favored practical jokes of the sort you set up and then retire from, much as one writes books, possibly because imagining people's reactions beat witnessing them. They procured a living hen and "hypnotized" it by setting it on the sink before the bathroom mirror in a friend's cottage by the Jersey shore. They spent weeks constructing a ten-foot sea monster—from truck inner tubes, cement blocks, broomsticks, lumber, pillows—and set it afloat in a friend's pond. On Sanibel Island, Florida, they baffled the shell collectors each Saint Patrick's Day by boiling a bucketful of fine shells in green dye and strewing the green shells up and down the beach before dawn. I woke one Christmas morning to find in my stocking, hung from the mantel with care, a leg. Mother had charmed a department store display manager into lending her one.

When I visited my friends, I was well advised to rise when their parents entered the room. When my friends visited me, they were well advised to duck.

WAKING UP WILD

WHAT DOES IT FEEL LIKE TO BE ALIVE? Living, you stand under a waterfall. You leave the sleeping shore deliberately; you shed your dusty clothes, pick your barefoot way over the high, slippery rocks, hold your breath, choose your footing, and step into the waterfall. The hard water pelts your skull, bangs in bits on your shoulders and arms. The strong water dashes down beside you and you feel it along your calves and thighs rising roughly back up, up to the roiling surface, full of bubbles that slide up your skin or break on you at full speed.

Can you breathe here? Here where the force is greatest and only the strength of your neck holds the river out of your face? Yes, you can breathe even here. You could learn to live like this. And you can, if you concentrate, even look out at the peaceful far bank where maples grow straight and their leaves lean down. For a joke you try to raise your arms. What a racket in your ears, what a scattershot pummeling!

It is time pounding at you, time. Knowing you are

alive is watching on every side your generation's short time falling away as fast as rivers drop through air, and feeling it hit.

Who turned on the lights? You did, by waking up: You flipped the light switch, started up the wind machine, kicked on the flywheel that spins the years. Can you catch hold of a treetop, or will you fly off the diving planet as she rolls? Can you ride out the big blow on the trunk of a coconut palm till the winds let up and you fall back asleep? You do, you fall asleep again, and you slide in a dream to the palm tree's base; the winds die off, the lights dim, the years slip away as you idle there till you die in your sleep, till death sets you cruising.

Knowing you are alive is feeling the planet buck under you—rear, kick, and try to throw you—while you hang on to the ring. It is riding the planet like a log downstream, whooping. Or, conversely, you step aside from the dreaming fast loud routine and feel time as a stillness about you, and hear the silent air ask in so thin a voice, Have you noticed yet that you will die? Do you remember, remember, remember? Then you may feel your life as a weekend, a weekend you cannot extend.

When I was fifteen, I felt it coming; now I was sixteen, and it hit.

My feet had been set imperceptibly on a new path, a fast path into a long tunnel like those many turnpike tunnels near Pittsburgh, turnpike tunnels whose entrances bear on brass plaques a roll call of the men who died blasting them. I wandered witlessly forward and found myself going down, and saw the light dimming; I adjusted to the slant and dimness, traveled farther down, adjusted to greater dimness, and so on. There wasn't a whole lot I could do about it, or about anything. I was going to hell on a handcart, that was all, and I knew it and everyone around me knew it, and there it was.

I was growing and thinning, as if pulled. I was getting angry, too, as if pushed. I morally disapproved of most things in North America, and blamed my innocent parents for them. My feelings deepened and lingered. The swift moods of early childhood—each formed by and suited to its occasion—vanished. Now feelings lasted so long they left stains. They arose from nowhere, like winds or waves, and battered me or engulfed me.

When I was angry, I felt myself coiled and longing to kill someone or bomb something big. One winter, trying to appease myself, I whipped my bed every afternoon with my uniform belt. I despised the spectacle I made in my own eyes—whipping the bed with a belt, like a creature demented!—and I often began halfheartedly, but I did it daily, after school, as a desperate discipline, trying to rid myself of my wildness.

Sometimes in class I couldn't stop laughing; things were too funny to be borne. It began then, my surprise that no one else saw what was so funny.

I read some few books with such reverence I didn't close them at the finish, but only moved the pile of pages back to the start, without breathing, and began again. I read one such book, an enormous novel, six times that way—closing the binding between sessions, but not between readings.

On the piano in the basement I played the maniacal "Poet and Peasant Overture" so loudly, for so many hours, night after night, I damaged the piano's keys and strings. When I wasn't playing this crashing overture, I played boogie-woogie, or something else, anything else, in octaves—otherwise, it wasn't loud enough. My fingers were so strong I could do push-ups with them. I played one piece with my fists. I banged on a steel-stringed guitar till I bled, and once, on a particularly great rock-and-roll downbeat, I broke straight through one of Father's snare drums.

I loved my boyfriend so tenderly, I thought I'd transmogrify into vapor. It would take spectroscopic analysis to locate my molecules in thin air. No way of holding him was close enough. Nothing could cure this bad case of gentleness except, perhaps, violence: if he swung me by the legs, maybe, and split my skull on a tree? Would that ease this insane wish to kiss too much his eyelids' outer corners and his temples, as if I could love up his brain?

I envied people in books who swooned. For two years I felt myself continuously swooning and continuously unable to swoon. The blood drained from my face and eyes and flooded my heart; my hands emptied, my knees unstrung, I bit at the air for something worth breathing—but I failed to fall, and I couldn't find the way to black out. I had to live on the lip of a waterfall, exhausted.

When I was bored I was first hungry, then nauseated, then furious and weak. "Calm yourself," people had been saying to me all my life. Since early childhood I had tried one thing and then another to calm myself, on those few occasions when I truly wanted to. Eating helped; singing helped. Now sometimes I truly wanted to calm myself. I couldn't lower my shoulders; they seemed to wrap around my ears. I couldn't lower my voice, although I could see the people around me flinch. I waved my arm in class till the teachers themselves wanted to kill me.

I was what they called a live wire. I was shooting out sparks that were digging a pit all around me, and I was sinking into that pit. Laughing with Ellin at school recess, or driving around after school with Judy in her jeep, exultant, or dancing with my boyfriend to Louis Armstrong across a polished dining-room floor, I got so excited I looked around wildly for aid. I didn't know where I should go or what I should do with myself. People in books split wood.

When rage reappeared, or boredom, it seemed never to have left. Each so filled me with so many years' intolerable accumulation, it jammed the space behind my eyes so I couldn't see. There was no room left to live. My rib cage was so tight I couldn't breathe. Every cubic centimeter of atmosphere above my shoulders and head was heaped with last straws. I couldn't peep, I couldn't wiggle or blink; my blood was too mad to flow.

For as long as I could remember, I had been transparent to myself, unself-conscious, learning, doing, most of every day. Now I was in my own way; I myself was a dark object I could not ignore. I couldn't remember how to forget myself. I didn't want to think about myself, to reckon myself in, to deal with myself every livelong minute on top of everything else—but swerve as I might, I couldn't avoid it. I was a boulder blocking my own path. I was a dog barking between my own ears, a barking dog who wouldn't hush.

So this was adolescence. Was this how the people around me had died on their feet—inevitably, helplessly? Perhaps their own selves eclipsed the sun for so many years the world shriveled around them, and when at last their inescapable orbits had passed through these dark egoistic years, it was too late; they had adjusted.

Must I then lose the world forever, that I had so loved? Was it all, the whole bright and various planet, where I had been so ardent about finding myself alive, only a passion peculiar to children, that I would outgrow even against my will?

OLD STONE PRESBYTERIAN

I QUIT THE CHURCH. I wrote the minister a fierce letter. The assistant minister, kindly Dr. James H. Blackwood, called me for an appointment. My mother happened to take the call.

"Why," she asked, "would he be calling you?" I was in the kitchen after school. Mother was leaning against the pantry door, drying a crystal bowl.

"What, Mama? Oh. Probably," I said, "because I wrote him a letter and quit the church."

"You—what?" She began to slither down the doorframe, weak-kneed, like Lucille Ball. I believe her whole life passed before her eyes.

As I climbed the stairs after dinner I heard her moan to Father, "She wrote the minister and quit the church."

"She—what?"

Father knocked on the door of my room. I was the only person in the house with a solitary room. Father ducked under the doorway, entered, and put his hands in his khakis' pockets.

"Hi, Daddy."

Actually, it drove me nuts when people came in my room. Mother had come in just last week. My room was getting to be quite the public arena. Pretty soon they'd put it on the streetcar routes. Why not hold the US Open here? I was on the bed, in uniform, trying to read a book. I sat up and folded my hands in my lap.

I knew that Mother had made him come—"She listens to you." He doubtless had been trying to read a book too.

Father looked around, but there wasn't much to see. My rock collection was no longer in evidence. A framed tiger swallowtail, slightly askew on its white cotton backing, hung on a yellowish wall. On the mirror I'd taped a pencil portrait of Rupert Brooke looking off softly. He looked like my boyfriend. Balanced on top of the mirror were some yellow-and-black FALLOUT SHELTER signs, big aluminum ones that my friends had collected as part of their antiwar efforts. On the pale maple desk there were, among other books and papers, an orange thesaurus, a blue three-ring binder with a boy's name written all over it in every typeface, a green assignment notebook, and Emerson's *Essays*.

Father began with some vigor: "What was it you

said in this brilliant letter?" At my silence he went on: Didn't I see? That people did these things—quietly? Just—quietly? No fuss? No flamboyant gestures. No uncalled-for letters. He was forced to conclude that I had deliberately set out to humiliate Mother and him.

"And your poor sisters too!" Mother added feelingly from the hall outside my closed door. She must have been passing at that very moment.

Just then we all heard a hideous shriek ending in a wail; it came from my sisters' bathroom. Had Molly cut off her head? It set us all back a moment—me on the bed, Father standing by my desk, Mother outside the closed door—until we all realized it was Amy, mad at her hair. Like me, Amy was undergoing a trying period, years long; she, on her part, was mad at her hair. She screeched at it wherever she was, the sound carrying all over the house, and it terrified all the rest of us, every time.

The assistant minister of the Shadyside Presbyterian Church, Dr. Blackwood, and I had a cordial meeting in his office. He was an experienced, calm man in a three-piece suit; he had a mustache and wore glasses. After he asked me why I had quit the church, he loaned me four volumes of C. S. Lewis's broadcast talks for a paper I was writing. Among the volumes proved to be *The Problem of Pain,* which I would find fascinating, not quite serious enough, and much too short. I had already written a paper on the Book of Job. The subject scarcely seemed to be closed. If the all-powerful creator directs the world, then why all this suffering? Why did the innocents die in the camps, and why do they starve in the cities and farms? Addressing this question, I found thirty pages written thousands of years ago, and forty pages written in 1955. They offered a choice of fancy language saying, "Forget it," or serenely worded, logical-sounding answers that so strained credibility (pain is God's megaphone) that "Forget it" seemed in comparison a fine answer.

TURNING OUT BADLY

FUNNY HOW BADLY I'D TURNED OUT. Now I was always in trouble. It felt as if I was doing just as I'd always done—I explored the neighborhood, turning over rocks. The latest rocks were difficult. I'd been in a drag race, of all things, the previous September, and in the subsequent collision, which landed me in the hospital; my parents saw my name in the newspapers, and their own names in the newspapers. Some boys I barely knew had cruised by that hot night and said to a clump of us girls on the sidewalk, "Anybody want to come along for a drag race?"

I did, absolutely. I loved fast driving.

It was then, in the days after the drag race, that I noticed the ground spinning beneath me, all bearings lost, and recognized as well that I had been loose like this—detached from all I saw and knowing nothing else—for months, maybe years. I whirled through the air like a bull-roarer spun by a lunatic who'd found his rhythm. The pressure almost split my skin. What else can you risk with all your might but your life?

Time unrolled before me in a line. I woke up and found myself in juvenile court. I was hanging from crutches; for a few weeks after the drag race, neither knee worked. In juvenile court, a policeman wet all ten of my fingertips on an inkpad and pressed them, one by one, using his own fingertips, on a form for files.

I'd already been suspended from school for smoking cigarettes. That was a month earlier, in early spring. Both my parents wept. Amy saw them, and began to cry herself. Molly, too, cried. She was six, missing her front teeth. Like the rest of us females, she had pale skin that turned turgid and red when she cried. She looked as if she were dying of wounds. I was the only one who didn't cry; but then, I was an intercontinental ballistic missile with an atomic warhead. They don't cry.

Late one night, my parents sat at the kitchen table; there was a truce. We were all helpless, and tired of fighting. Amy and Molly were asleep.

"What are we going to do with you?" Mother raised

the question. Her voice trembled and rose with emotion.

She couldn't sit still; she kept getting up and roaming around the kitchen. Father stuck out his chin and rubbed it with his big hands. I covered my eyes. Mother squeezed white lotion into her hands, over and over. We all smoked; the ashtray was full. Mother walked over to the sink, poured herself some ginger ale, ran both hands through her short blond hair to keep it back, and shook her head.

She sighed and said again, looking up and out of the night-black window, "Dear God, what are we going to do with you?"

My heart went out to them. We all seemed to have exhausted our options. They asked me for fresh ideas, but I had none. I racked my brain, but couldn't come up with anything. The US Marines didn't take sixteen-year-old girls.

ENVOY

OUR FATHER TAUGHT US THE CULTURE into which we were born. American culture was Dixieland above all, Dixieland pure and simple, and next to Dixieland, jazz. It was the pioneers who went west singing "Bang away, my Lulu." When someone died on the Oregon Trail, as someone was always doing, the family scratched a shallow grave right by the trail, because the wagon train couldn't wait. Everyone paced on behind the oxen across the empty desert and some families sang "Bang away, my Lulu" that night, and some didn't.

Our culture was the stock-market crash—the biggest and best crash a country ever had. Father explained the mechanics of the crash to young Amy and me, around the dining-room table. He tried to explain why men on Wall Street had jumped from skyscrapers when the stock market crashed: "They lost everything!"—but of course I thought they lost everything only when they jumped. It was the breadlines of the Depression, and the Okies fleeing the Dust Bowl, and the proud men begging on city streets, and families on the move seeking work—dusty women, men in black hats pulled over their eyes, haunted, hungry children: what a mystifying spectacle, this almost universal misery, city families living in cars, farm families eating insects, because—why? Because all the businessmen realized at once, on the same morning, that paper money was only paper. What terrible fools. What did they think it was?

American culture was the World's Fair in Chicago, baseball, the Erie Canal, fancy nightclubs in Harlem, silent movies, summer-stock theater, the California forty-niners, the Alaska gold rush, Henry Ford and his bright idea of paying workers enough to buy cars, P. T. Barnum and his traveling circus, Buffalo Bill Cody and his Wild West Show. It was the Chrysler Building in New York and the Golden Gate Bridge in San Francisco; the *Monitor* and the *Merrimack*, the Alamo, the Little Bighorn, Gettysburg, Shiloh, Bull Run, and "Strike the tent."

It was Pittsburgh's legendary Joe Magarac, the mighty Hungarian steelworker, who took off his shirt to reveal his body made of high-grade steel, and who squeezed out steel rail between his knuckles by the ton. It was the brawling rivermen on the Ohio River, the sandhogs who dug Hudson River tunnels, silver miners in Idaho, cowboys in Texas, and the innocent American Indian Jim

Thorpe, who had to give all his Olympic gold medals back. It was the men of every race who built the railroads, and the boys of every race who went to war.

Above all, it was the man who wandered unencumbered by family ties: Johnny Appleseed in our home woods, Daniel Boone in Kentucky, Jim Bridger crossing the Rockies. Father described for us the Yankee peddler, the free trapper, the roaming cowhand, the whaler man, roustabout, gandy dancer, tramp. His heroes, and my heroes, were Raymond Chandler's city detective Marlowe going, as a man must, down these mean streets; Huck Finn lighting out for the territories; and Jack Kerouac on the road.

Every time we danced, Father brought up Jack Kerouac, *On the Road*.

We did a lot of dancing at our house, fast dancing; everyone in the family was a dancing fool. I always came down from my room to dance. When the music was going, who could resist? I bounced down the stairs to the rhythm and began to whistle a bit, helpless as a marionette whose strings jerk her head and feet.

We danced by the record player in the dining room. For fast dancing, Mother only rarely joined in; perhaps Amy, Molly, and I had made her self-conscious. We waved our arms a lot. I bumped into people, because I liked to close my eyes.

"Turn that record player down!" Mother suggested from the living room. She was embroidering a pillow. Father opened the cabinet and turned the volume down a bit. I opened my eyes.

"Remember that line in *On the Road*?" He addressed me, because between us we had read *On the Road* approximately a million times. Like *Life on the Mississippi,* it was the sort of thing we read. I thought of his blue bookplate: "Books make the man." The bookplate's ship struggled in steep seas, and crowded on too much sail.

I nodded; I knew what he was going to say, because he said it every time we played music; it was always a pleasure. We both reined in our dancing a bit, so we could converse. Sure I remembered that line in *On the Road*.

"Kerouac's in a little bar in Mexico. He says that was the only time he ever got to hear music played loud enough—in that little bar in Mexico. It was in *On the Road*. The only time he ever got to hear the music loud enough. I always remember that." He laughed, shaking his head; he turned the record player down another notch.

It had been a long time since Father had heard the music played loud enough. Maybe he was still imagining it, fondly, some little bar back away somewhere, so small he and the other regulars sat in the middle of the blaring band, or stood snapping their fingers, drinking bourbon, telling jokes between sets. He knew a lot of jokes. Did he think of himself as I thought of him, as the man who had cut out of town and headed, wearing tennis shoes and a blue cap, down the river toward New Orleans?

I was gaining momentum. It was only a matter of months till I went to college and got free. Downstairs in the basement, I played "Shake, Rattle, and Roll" on the piano. Why not take up the trumpet, why not marry this wonderful boy, write an epic, run a medical mission in the Amazon as always intended? What happened to painting, what happened to science? My boyfriend seemed never to sleep. "I can sleep when I am dead," he said. Was this not grand?

I was approaching escape velocity. What would you do if you had fifteen minutes to live before the bomb went off? Quick: What would you read?

I drove up and down the boulevards—fast—up and down the highways, around Frick Park, over the flung bridges, and up into the springtime hills. My boyfriend and I played lightning chess, ten games an hour. We drove up the Allegheny River into West Virginia and back. In my room I shuffled cards. I wrote poems about the sea. I wrote poems imitating the psalms. I held my pen on the red paper label of the modern jazz record on the turntable, played that side past midnight over and over, and let the pen draw a circle an hour thick.

In New Orleans—if you could get to New Orleans—would the music be loud enough?

A WRITER IN THE WORLD

PEOPLE LOVE PRETTY MUCH THE SAME THINGS BEST. A writer, though, looking for subjects asks not after what he loves best, but what he alone loves at all. Strange seizures best us. Frank Conroy loved his yo-yo tricks, Emily Dickinson her slant of light; Richard Selzer loves the glistening peritoneum; Faulkner, the muddy bottom of a little girl's drawers just visible when she's up a pear tree. "Each student of the ferns," I read, "will have his own list of plants that for some reason or another stir his emotions."

Why do you never find anything written about that idiosyncratic thought you advert to, about your fascination with something no one else understands? Because it is up to you. There is something you find interesting, for a reason hard to explain because you have never read it on any page; there you begin. You were made and set here to give voice to this, your own astonishment. "The most demanding part of living a lifetime as an artist is the strict discipline of forcing oneself to work steadfastly along the nerve of one's own most intimate sensitivity." Anne Truitt, the sculptor, said this. Thoreau said it another way: Know your own bone. "Pursue, keep up with, circle round and round your life. . . . Know your own bone: gnaw at it, bury it, unearth it, and gnaw at it still."

Write as if you were dying. At the same time, assume you write for an audience consisting solely of terminal patients. That is, after all, the case. What would you begin writing if you knew you would die soon? What could you say to a dying person that would not enrage by its triviality?

Write about winter in the summer. Describe Norway as Ibsen did, from a desk in Italy; describe Dublin as James Joyce did, from a desk in Paris. Willa Cather wrote her prairie novels in New York City; Mark Twain wrote *Huckleberry Finn* in Hartford, Connecticut. Recently, scholars learned that Walt Whitman rarely left his room.

The writer studies literature, not the world. He lives in the world; he cannot miss it. If he has ever bought a hamburger, or taken a commercial airplane flight, he spares his readers a report of his experience. He is careful of what he reads, for that is what he will write. He is careful of what he learns, because that is what he will know.

The writer as a consequence reads outside his time and place. He reads great novels: *Daniel Deronda,* say, and the novels of James McBride. His nonconformity may be his only hope.

The writer knows his field—what has been done, what could be done, the limits—the way a tennis player knows the court. And he, too, plays the edges. That's where the exhilaration is: He hits up the line. He pushes the edges. Beyond this limit, here, the reader must recoil. Reason balks, poetry snaps, some madness enters, or strain. Now courageously and carefully, can he enlarge it? Can he nudge the bounds? And enclose what wild power?

The body of literature, with its limits and edges, exists outside some people and inside others. Only after the writer lets literature shape her can she perhaps shape literature. In working-class France, when an apprentice got hurt, or when he got tired, the experienced workers said, "It is the trade entering his body." The art must enter the body too. A painter cannot use paint like glue or screws to fasten down the world. The tubes of paint are like fingers; they work only if, inside the painter, the neural pathways are wide and clear to the brain. Cell by cell, molecule by molecule, atom by atom, part of the brain changes physical shape to fit the paint.

You adapt yourself, Paul Klee said, to the contents of the paintbox. Adapting yourself to the contents of the paintbox, he said, is more important than nature and its study. The painter, in other words, does not fit the paints to the world. He most certainly does not fit the world to himself. He fits himself to the paint. The self is the servant who bears the paintbox and its inherited contents. Klee called this insight, quite rightly, "an altogether revolutionary new discovery."

A well-known writer got collared by a university student who asked, "Do you think I could be a writer?"

"Well," the writer said, "I don't know. . . . Do you like sentences?"

The writer could see the student's amazement. Sentences? Do I like sentences? I am twenty years old, and do I like sentences?

If he had liked sentences, of course, he could have begun, like a joyful painter I knew. I asked him how he came to be a painter. He said, "I liked the smell of the paint."

Hemingway studied, as models, the novels of Knut Hamsun and Ivan Turgenev. Isaac Bashevis Singer, as it happened, also chose Hamsun and Turgenev as models. Ralph Ellison studied Hemingway and Gertrude Stein. Thoreau loved Homer; Eudora Welty loved Chekhov. Faulkner described his debt to Sherwood Anderson and Joyce; E. M. Forster, his debt to Jane Austen and Proust. By contrast, if you ask a twenty-one-year-old poet whose poetry he likes, he might say, unblushing, "Nobody's." In his youth, he has not yet understood that poets like poetry, and novelists like novels; he himself likes only the role, the thought of himself in a hat.

Rembrandt and Shakespeare, Tolstoy and Gauguin, possessed, I believe, powerful hearts, not powerful wills. They loved the range of material they used, the work's possibilities excited them; the field's complexities fired their imaginations. The caring suggested the tasks; the tasks suggested the schedules. They learned their fields and then loved them. They worked, respectfully, out of their love and knowledge, and they produced complex bodies of work that endure. Then, and only then, the world maybe flapped at them some sort of hat, which, if they were still living, they ignored as well as they could, to keep at their tasks.

It makes more sense to write one big book—a novel or nonfiction narrative—than to write many stories or essays. Into a long, ambitious project you can fit or pour all you possess and learn. A project that takes five years will accumulate those years' inventions and richnesses. Much of those years' reading will feed the work. Further, writing sentences is difficult whatever their subject. It is no less difficult to write sentences in a recipe than sentences in *Moby-Dick*. So you might as well write *Moby-Dick*. Similarly, since every original work requires a unique form, it is more prudent to struggle with the outcomes of only one form—that of a long work—than to struggle with the many forms comprising a collection. Each chapter of a prolonged narrative is problematic, too, of course, and the writer undergoes trials as the structure collapses and coheres by turns—but at least the labor is not all on spec. The chapter already has a context: a tone, setting, characters. The work is already off the ground. You must carry the reader along, of course, but you need not, after the first chapters, bear him aloft while performing a series of tricky introductions.

Writing every book, the writer must solve two problems: Can it be done? and, Can I do it? Every book has an intrinsic impossibility that its writer discovers as soon as his first excitement fades. The problem is structural; it is insoluble; it is why no one can ever write this book. Complex stories, essays, and poems have this problem, too—the prohibitive structural defect the writer wishes he'd never noticed. He writes it in spite of that. He finds ways to minimize the difficulty; he strengthens other virtues; he cantilevers the whole narrative out into thin air, and it holds. If it can be done, then he can do it, and only he. For there is nothing in this material that suggests to anyone but him its possibilities for meaning and feeling.

Why are we reading, if not in hope of beauty laid bare, life heightened, and its deepest mystery probed? Can the writer isolate and vivify all in experience that most deeply engages his intellect and heart—and our own? Can the writer renew our hope for literary forms? Why are we reading if not in hope that the writer will magnify and dramatize our days, will illuminate and inspire us with wisdom, courage, and the possibility of meaning, and will press upon our minds the deepest mysteries, so we may feel again their majesty and power? What do we ever know that is higher than the power which, from time to time, seizes our lives and reveals us startlingly to ourselves as creatures set down here bewildered? Why does death so catch us by surprise, and why love? We still and always want waking. We should amass half-dressed in long lines like tribesmen and shake gourds at one another, to wake up; instead we watch television and miss the show.

And if we are reading for these things—and only if—why would any of us read books with advertising slogans and brand names in them? Why would anyone write such books? Commercial intrusion has overrun and crushed, like the last glaciation, a humane landscape. The new landscape and its climate put metaphysics on the run. Must writers collaborate? Well, in fact, the novel as a form has only rarely been metaphysical; usually it presents society as it is. The novel often aims to fasten down the spirit of its time, to make a heightened simulacrum of our recognizable world in order to present it shaped and analyzed. This has never seemed to me worth doing, but it is certainly one thing literature has always done. Writers attracted to metaphysics can simply ignore the commercial blare, as if it were a radio, or use historical settings, or flee to nonfiction or poetry.

The sensation of writing a book is the sensation of spinning, blinded by love and daring. It is the sensation of rearing and peering from the bent tip of a grass-blade, looking for a route. At its absurd worst, it feels like what mad Jacob Boehme, the German mystic, described in his first book. He was writing incoherently as usual, about the source of evil. The passage here, though, will serve as well for the source of books.

> The whole Deity has in its innermost or beginning Birth, the Pith or Kernel, a very tart, terrible *Sharpness,* in which the astringent Quality is a very horrible, tart, hard, dark and cold Attraction or Drawing together, like *Winter,* when there is a fierce, bitter cold Frost, when Water is frozen into Ice, and besides is very intolerable.

If you can dissect out the very intolerable, tart, hard, terribly sharp Pith or Kernel, and begin writing the book compressed therein, the sensation changes. Now it feels like alligator wrestling, at the level of the sentence. This is your life. You are a Seminole alligator wrestler. Half-naked, with your two bare hands, you hold and fight a sentence's head while its tail tries to knock you over.

Several years ago in Florida, an alligator wrestler lost. He was grappling with an alligator in a lagoon in front of a paying crowd. The crowd watched the young Seminole and the alligator twist belly to belly in and out of the water; after one plunge, they failed to rise. A writer named Lorne Ladner described it. Bubbles came up on the water. Then blood came up, and the water stilled. As the minutes elapsed, the people in the crowd exchanged glances; silent, helpless, they quit the stands. It took the Seminoles a week to find the man's remains.

At its best, the sensation of writing is that of any unmerited grace. It is handed to you, but only if you look for it. You search, you break your heart, your back, your brain, and then—and only then—it is handed to you. From the corner of your eye you see motion. Something is moving through the air and headed your way, on two white wings. It flies directly at you; you can read your name on it. If it were a baseball, you'd hit it out of the park. It is that one pitch in a thousand you see in slow motion; its wings beat slowly as a hawk's.

One line of a sonnet, the poet said—only one line of fourteen, but thank God for that one line—drops from the ceiling.

Push it. Examine all things intensely and relentlessly. Probe and search each object in a piece of art. Do not leave it, do not course over it as if it were understood, but instead follow it down until you see it in the mystery of its own specificity and strength. Giacometti's drawings and paintings show both his bewilderment and persistence. If he had not acknowledged his bewilderment, he would not have persisted. A twentieth-century master of drawing, Rico Lebrun, taught that "the draftsman must aggress; only by persistent assault will the live image capitulate and give up its secret to an unrelenting line." Who but an artist fierce to know—not fierce to *seem* to know—would suppose that a live image possessed a secret? The artist is willing to give all his or her strength and life to probing with blunt instruments those same secrets no one can describe in any way but with those instruments' faint tracks.

Admire the world for never ending on you—as you would admire an opponent, without taking your eyes from him, or walking away.

One of the few things I know about writing is this: Spend it all, shoot it, play it, lose it, all, right away, every time. Don't hoard what seems good for a later place in the book, or for another book; give it, give it all, give it now. The very impulse to save something good for a better place later is the signal to spend it now. Something more will arise for later, something better. These things fill from behind, from beneath, like well water. Similarly, the impulse to keep to yourself what you have learned is not only shameful; it is destructive. Anything you do not give freely and abundantly becomes lost to you. You open your safe and find ashes.

After Michelangelo died, someone found in his studio a piece of paper on which he had written a note to his apprentice, in the handwriting of his old age: "Draw, Antonio, draw, Antonio, draw and do not waste time."

THIS IS THE LIFE

ANY CULTURE TELLS YOU HOW TO LIVE your one and only life: to wit, as everyone else does.

Probably most cultures prize, as ours rightly does, making a contribution by working hard at work that you love; being in the know, and intelligent; gathering a surplus; and loving your family above all, and your dog, your cat; bird-watching. Beyond those things our culture may place a special focus on money, on celebrity, on physical beauty. These are not universal.

Elsewhere it might be: You wear the best shoes you can afford, you seek to know Rome's best restaurants and their staffs, drive the best car, and vacation on Tenerife. And what a cook you are!

Or you take the next tribe's pigs in thrilling raids; you grill yams; you trade for televisions and hunt white-plumed birds. Everyone you know agrees: This is the life. Perhaps you burn captives. You set fire to drunks. Yours is the human struggle, or the elite one, to achieve . . . whatever your own culture tells you: to publish the paper that proves the point; to progress in the firm and gain high title and salary, stock options, benefits; to get the loan to store the beans till their price rises; to elude capture; to feed your children or educate them to a feather edge; to count coup or perfect your calligraphy; to eat the king's deer or catch the poacher; to spear the seal, intimidate the enemy, and be a big man or beloved woman and die respected for the pigs or the title or the shoes. Not a funeral. Forget funeral. A big birthday party.

Since everyone around you agrees ever since there were people on earth that land is value, or labor is value, or learning is value, or title, degree, necklaces, murex shells, the ownership of slaves. Everyone knows bees sting and ghosts haunt and giving your robes away humiliates your rivals. That the enemies are barbarians. That wise men swim through the rock of the earth; that houses breed filth, airstrips attract airplanes, tornadoes punish, ancestors watch, and you can buy a shorter stay in purgatory. The black rock is holy, or the scroll; or the pangolin is holy, the quetzal is holy, this tree, water, rock, stone, cow, cross, or mountain—and it's all true. The Red Sox. Or nothing at all is holy, as everyone intelligent knows.

Who is your everyone? Chess masters scarcely surround themselves with motocross racers. Do you want aborigines at your birthday party? Or is it yak-butter tea you are serving? Popular culture deals not in its distant past, or any other past, or any other culture. You know no one who longs to buy a mule or be presented at court or thrown into a volcano.

So the illusion, like the visual field, is complete. It has no holes except books you read and soon forget. And death takes us by storm. What was that, that life? What else offered? If for him it was contract bridge, if for her it was copyright law, if for everyone it was and is an optimal mix of family and friends, learning, contribution, and joy—of making and ameliorating—what else is there, or was there, or will there ever be?

What else is a vision or fact of time and the peoples it bears issuing from the mouth of the cosmos, from the round mouth of eternity, in a wide and parti-colored utterance. In the complex weave of this utterance like fabric, in its infinite domestic interstices, the centuries and continents and classes dwell. Each people knows only its own squares in the weave, its wars and instruments and arts, and also perhaps the starry sky.

Okay, and then what? Say you manage to scale your own weft and see time's breadth and the length of space. You see the way the fabric both passes among the stars and encloses them. You see in the weave nearby, and aslant farther off, the peoples variously scandalized or exalted in their squares. They work on their projects—they flake spear points, hoe, plant; they kill aurochs or one another; they prepare sacrifices—as we here and now work on our project. What, seeing this spread multiply infinitely in every direction, would you do differently? Would you change your project? To what? Whatever you do, it has likely brought delight to fewer people than either contract bridge or the Red Sox.

However hypnotized you and your people are, you will be just as dead in their war, our war. However dead you are, more people will come. However many more people come, your time and its passions, you yourself and your passions, weigh but equally in the balance with those of any dead who pulled waterwheel poles by the Nile or Yellow Rivers, or painted their foreheads black, or starved in the wilderness, or wasted from disease, then or now. Our lives and our deaths surely count equally, *or* we must abandon one-man-one-vote, dismantle democracy, and assign seven billion people an importance-of-life ranking from one to seven billion.

What would you do differently, you up on your beanstalk looking at scenes of all peoples at all times in all places? When you climb down, would you dance any less to the music you love, knowing that music to be provisional as a bug? Somebody has to make jogging shoes, to turn the soil, fish. If you descend the long rope ladders back to your people, your own time in the fabric, if you tell them what you have seen, and should someone care to listen, then what? Everyone knows times and cultures are plural. If you come back a shrugging relativist or a stiff-tongued absolutist, then what? If you spend hours a day looking around, high astraddle the warp or woof of your people's wall, then what new wisdom might you take to your grave for worms to untangle? Well, maybe you will not go into advertising. But what work suits? You might know your own death better, though dread it no less. Will you try to bring people up the wall—carry children to see it—to what end? Fewer golf courses? What's wrong with golf? Nothing at all. Equality of wealth? Sure; how?

The woman watching sheep over there, the man who carries embers in a pierced clay ball, the engineer, the girl who spins wool into yarn as she climbs, the smelter, the babies learning to recognize speech in their own languages, the man whipping a slave's flayed back, the man digging roots, the woman digging roots, the child digging roots—what would you tell them? And the future people—what are they doing? What excitements sweep peoples here and there from time to time? Into the muddy river they go, into the trenches, into the caves, into the mines, into the granary, into the sea in boats. Most humans who were ever alive lived inside a single culture that had not changed for hundreds of thousands of years.

Over here, the rains fail; they are starving. There, the caribou fail; they are starving. Corrupt leaders take the wealth. Not just there, but here. Rust and smut spoil the rye. When pigs and cattle starve or freeze, people die soon after. Disease empties a sector, a billion sectors.

People look at the sky and at the other animals. They make beautiful objects, beautiful sounds, beautiful motions of their bodies beating drums in lines. They pray; they toss people in peat bogs; they help the sick and injured; they pierce their lips, their noses, ears; they make the same mistakes despite religion, written language, philosophy, and science. They build, they kill, they preserve, they count and figure, they boil the pot, they keep the embers alive; they tell their stories and gird themselves.

Will knowledge you experience directly make you a Buddhist? Must you forfeit excitement per se? To what end?

Say you have seen an ordinary bit of what is real, the infinite fabric of time that eternity shoots through, and time's soft-skinned people working and dying under slowly shifting stars. Then what?

DISNEYLAND

IT IS A SUNNY SEPTEMBER MORNING in Disneyland. Bands are playing; people walk with their children and take pictures.

The Chinese writers, the UCLA Chinese-American writers' conference hosts, and Allen Ginsberg and I—for we stayed on to accompany our foreign guests—have just seen the film *America the Beautiful.* Along with the standard scenic tourist shots, the film offered a healthy dose of American militarism. Tanks rolled on parade, soldiers fired salutes, cadets trained with weapons at Annapolis and West Point—all to swelling music and rising choruses.

We have emerged, blinking, from this film and entered again the bright Disneyland streets. The Chinese writers seem content, perfectly familiar with Disney paraphernalia. In China you can buy Donald Duck on pink thermos bottles, Mickey Mouse and Goofy on handkerchiefs. Nearly everyone has seen Disney cartoons.

A sophisticated and cosmopolitan Chinese writer named Liu Binyan is strolling down the street with Allen Ginsberg. At home in Beijing, Liu Binyan is a muckraking journalist. He revealed corruptions in high places. He is in the United States on a six-month visit, he says. In fact, this visit will prove to be the start of his lifelong exile from China. The party disliked corruption as a topic. Liu Binyan's English is perfect. And his upright, forceful carriage enhances the grandeur of his leonine head, with its curved forehead, wide cheekbones, and strong jaw. He is young, at home in the world; his dark suit, remarkably, fits him. For twenty-two years in China, no longer permitted to write, he worked at forced labor. Now, he is in Disneyland.

Allen Ginsberg, beside Liu Binyan, is walking with his head down. He is sensibly dressed for a hot September day in a white, short-sleeved shirt and green chinos. The spectacle of the movie we just screened has left him gloomy. He says he considers all that military emphasis in the film to be Mickey Mouse.

Liu Binyan, walking so erectly in his fine suit, cocks an ear and says, "Mickey Mouse?"

"You know," Ginsberg says. He is preoccupied. "Mickey Mouse. With the ears?" He wags his fingers desolately over his head. "A little mouse?"

Liu Binyan stands on his dignity. "Yes," he says

slowly, in his careful English, "I know Mickey Mouse. Yes. But the film?"

Ginsberg is emphatic. "That was a Mickey Mouse film."

It is all breaking down for Liu Binyan. He has probably seen dozens of Mickey Mouse films. Incredulity raises his voice: "The film we just saw was a Mickey Mouse film?"

Ginsberg, still shaking his head over the film, chooses another tack. "You know," he explains. "Hallucinatory. Delusional."

Liu Binyan slowly lights a cigarette and lets the subject go.

We all come around a corner and a band is playing. We are alone on a broad intersection under blue eucalyptus trees that cast pale and wobbling lines of shadow on the street. Two of us Americans begin to dance.

One of the Chinese men, with debonair smoothness, as if this were what one did every lunch hour on the streets of Beijing, lightly taps one of the women, and they dance. The band is playing Duke Ellington—"Mood Indigo." They dance lightly, formally, grandly, seriously, until the song is over; we all continue on, without comment.

Things are always jolly when people misunderstand one another. I have made ludicrous mistakes in many places, in a number of the many languages I do not speak. In China people often looked at me aghast when I tried to pronounce a simple hello; I have no idea what I was actually saying. Several of the Chinese writers we are with, on the other hand, speak excellent English. Still, people will misunderstand one another.

We are luncheon guests at Disneyland's "Club 33." We have a private dining room, with access to a lavish and excellent buffet. With us are three Disneyland hostesses, each dressed in a red tailored suit. The hostess at my table, to my right, is a gracious and shining young woman whose enormous name tag reads SUSI. Her blond hair is blunt cut and curled in a flip. She has a wide, friendly smile; she has worked as a hostess in Disneyland for seven years.

To my left sits a Chinese official, one Mr. Fu, whose English is very good; we have all often relied on this man's patient good nature and perfect sense of propriety to ease conversation along. Just now he seems to be engaged in catching his breath a bit, and in parsing the edibles on his plate. The Chinese are apt to take eating rather more seriously than we do, and they are not so given as we to social chatter at table.

SUSI, having ascertained the excellence of her lunch companion's English, opens the luncheon conversation. "Well, Mr. Fu! Are you enjoying your visit to Disneyland?"

Mr. Fu looks up from his plate and smiles. "No," he says.

SUSI freezes. Fu goes back to his plate, spears a smoked oyster on his fork, and adds conversationally, "This is my very first visit."

Ah, a simple misunderstanding. Easily explained to everyone. After that is straightened out, we all resume eating with a certain concentration. Following a decent interval, SUSI, with commendable pluck, gives it another go.

"Well, Mr. Fu!"

Fu looks up brightly, as if eager to reestablish himself in the good graces of his hostess.

"How do you find Disneyland?"

Fu smiles broadly, raises his eyebrows for emphasis, and says clearly, quite pleasantly, "I find it very messy."

"Messy?" SUSI's smile has fallen, along with her fork.

"Messy," Fu repeats, alarmed. "A'messy. Yes? A'*mess*-ing."

Ah!! . . .

Seldom have I enjoyed a luncheon more.

Accompanying the delegation as its "secretary" is a middle-aged woman named Fan Baoci—"Madame Fan." She speaks English, rooms with Zhang Jie in every hotel, and, I believe, enjoys herself. She is a tidy, small woman in glasses, and seems a sober, responsible person. I am charmed to hear about her unexpectedly spirited home life: "My husband loves children, and he loves *me* of course! And he has a good sense of humor. He plays the piano, and the guitar, and the accordion. Often at my house in the evening, after work, we start to dance."

Now in Disneyland Madame Fan and I are alone on the street. The restaurant has chilled her a bit, and we are warming her in the sun. People are passing along the narrow street, which is modeled after a street in New Orleans's French Quarter.

A little boy approaches us. He is about five years old, a blond little boy with long bangs. He is wearing blue shorts, a green T-shirt, and a gun belt with two holsters. Madame Fan leans over and addresses him warmly: "What's your name?"

The little boy draws his two six-shooters and, with a grim "I-hate-to-do-this" expression, shoots us, one and then the other: Pow! Pow! I keel over, so he concentrates on Madame Fan: Pow!

"What's your name?" says Madame Fan, leaning down along the line of fire.

The boy steps back and draws a bead on her forehead: Pow pow pow! This goes on. I suggest to Madame Fan that she clutch at her breast and die a bit, in the interest of goodwill, and if she wants the killing to stop; and so she does. The boy pops his guns back into their holsters and stalks on down the street without a word.

The head of the delegation is a literary critic named Feng Mu. He is a rather shy, formal bachelor of sixty-four. He has an impressive gift for giving beautifully structured complimentary speeches off the cuff. Like the other writers, who perhaps take their cue from him, he seems game for any strange thing the United States may throw at him.

Feng Mu is so formal in his bearing that he manages, at all times, to stand with his spine at once perfectly aligned and canted backwards, away from the world, like a raked mast. Because of this posture, Feng Mu, who is not tall, nevertheless looks down his nose at things, and seems somewhat taken aback. Among his own delegation he is a favorite, not only for his personal qualities, but for the relative liberality of his politico-literary criticism and the honor it has brought him.

Among us Americans he is a favorite as well for a quality in his formality that I can only call sweetness.

Feng Mu and some others try the wildest ride at Disneyland: "Space Mountain." It is like a roller coaster, only its cars, instead of dropping, jerk and veer through hairpin turns in the dark. When it is all over, Feng Mu and the others climb from their cars breathless, and compose themselves on terra firma. Feng Mu pats a hand over his hair and restores it to order. He stands absolutely straight and tilted backwards; his legs are formally together; his expression is at once serene and exalted. "I think," he says severely, as if addressing the press, "that unless one has ridden 'Space Mountain,' one cannot truly claim to have been to Disneyland."

At some point we have managed to lose one of the Chinese. Chen Baichen is seventy-four years old and speaks no English. He must have been missing for quite a while before anyone noticed; in the meantime, we have taken a train ride across the park.

Chen Baichen is a playwright whose plays came out in the thirties and forties. He is a short, dignified man, broadly built, with large features and a long chin that touches his top collar button. Usually he wears, as do most of the Chinese, a rather long-sleeved trench coat.

Apparently their briefings stressed American rains.

This morning Chen Baichen and I were alone together briefly at the breakfast table in our hotel. While we waited for our scrambled eggs, he got up to leave for a minute. (He was fetching, it turned out, one of his books for a present.) Politely, he explained in Chinese that he was leaving for just a minute and would be right back. I understood, I said; fine, sure—for all this was evident—but then he remembered that I don't understand Chinese, and took this to mean that I didn't understand him.

So, quickly, leaning over his place mat, he did what Chinese people are constantly doing during verbal misunderstandings: He sketched the character with a finger. On his place mat he drew the Chinese characters for, I guess, "I'm just going upstairs for a minute and will be right back." In the middle of this exercise he awoke to its absurdity. I had been, in spite of myself, following the Chinese characters' forms raptly, but now he quickly wiped away the imaginary characters with his palm, made a wonderfully disgusted "Bah!" gesture with both hands, and left. I like everything about Chen Baichen, although I cannot claim to know him, and think it careless of us to have lost him.

A disagreement ensues in Disneyland: Are we to assume that Chen Baichen, wherever he is, is frantic, or at least upset? Some people think that Chen Baichen, having been through two world wars, occupation, liberation, famine, the anti-rightist campaign, and the Cultural Revolution, can probably handle Disneyland. In fact, as we learn later, he has calmly made his way to the park's exit and is waiting on a bench.

I have been assigned to comb a section of the park. By the time I meet up with the group, Chen Baichen has been "found" for twenty minutes. Nevertheless I am so happy to see him that I forget all the warnings in the guidebooks and hug him rather enthusiastically. As we part, I see, disheartened, that his enormous eyes are full of tears.

So he *had* been upset to be lost in Disneyland. I was wrong. Of course he had been upset, and was now relieved—who wouldn't be?

Later, however, I learn that Chen Baichen, on his own account, was not in the least ruffled by being lost in Disneyland. But the warmth of our relief and embraces when we joined him—that had moved him to tears. Chinese men lose no face by weeping.

ON FOOT IN VIRGINIA'S ROANOKE VALLEY

I USED TO HAVE A CAT, an old fighting tom, who would jump through the open window by my bed in the middle of the night and land on my chest. I'd half awaken. He'd stick his skull under my nose and purr, stinking of urine and blood. Some nights he kneaded my bare chest with his front paws, powerfully, arching his back, as if sharpening his claws, or pummeling a mother for milk. Some mornings I'd wake in daylight to find my body covered with paw prints in blood; it looked as though I'd been painted with roses.

It was hot, so hot the mirror felt warm. I washed before the mirror in a daze, my twisted summer sleep still hung about me like sea kelp. What blood was this, and what roses? It could have been the rose of union, the blood of murder, or the rose of beauty bare and the blood of some unspeakable sacrifice or birth. The sign on my body could have been an emblem or a stain, the keys to the kingdom or the mark of Cain. I never knew.

I never knew as I washed, and the blood streaked, faded, and finally disappeared, whether I'd purified myself or ruined the blood sign of the Passover. We wake, if we ever wake at all, to mystery, rumors of death, beauty, violence. . . . "Seem like we're just set down here," a woman said to me recently, "and don't nobody know why."

These are morning matters, pictures you dream as the final wave heaves you up on the sand to the bright light and drying air. You remember pressure, and a curved sleep you rested against, soft, like a scallop in its shell. But the air hardens your skin; you stand; leave the lighted shore to explore some dim headland, and soon you're lost in the leafy interior, intent, remembering nothing.

I still think of that old tomcat, mornings, when I wake. Things are tamer now; I sleep with the window shut. The cat and our rites are gone and my life is changed, but the memory remains of something powerful playing over me. I wake expectant, hoping to see a new thing. If I'm lucky I might be jogged awake by a strange birdcall. I dress in a hurry, imagining the yard flapping with auks, or flamingos. This morning it was a wood duck, down at the creek. It flew away.

I live by a creek, Tinker Creek, in a valley in Virginia's Blue Ridge. It's where I make myself scarce. An anchorite's hermitage is called an anchor-hold; some anchor-holds were simple sheds clamped to the side of a church like a barnacle to a rock. I think of this house clamped to the side of Tinker Creek as an anchor-hold. It holds me at anchor to the rock bottom of the creek itself and it keeps me steadied in the current, as a sea anchor does, facing the stream of light pouring down. It's a good place to live; there's a lot to think about.

The creeks—Tinker and Carvin's—are an active mystery, fresh every minute. Theirs is the mystery of the continuous creation and all that providence implies: the uncertainty of vision, the horror of the fixed, the dissolution of the present, the intricacy of beauty, the pressure of fecundity, the elusiveness of the free, and the flawed nature of perfection. The mountains—Tinker and Brushy, McAfee's Knob and Dead Man—are a passive mystery, the oldest of all. Theirs is the one simple mystery of creation from nothing, of matter itself, anything at all, the given. Mountains are giant, restful, absorbent. You can heave your spirit into a mountain and the mountain will keep it, folded, and not throw it back as some creeks will. The creeks are the world with all its stimulus and beauty; I live there. But the mountains are home.

The wood duck flew away. I caught only a glimpse of something like a bright torpedo that blasted the leaves where it flew. Back at the house I eat a bowl of oatmeal; much later in the day will come the long slant of light that means good walking.

If the day is fine, any walk will do; it all looks good. Water in particular looks its best, reflecting blue sky in the flat, and chopping it into graveled shallows and white chute and foam in the riffles. On a dark day, or a hazy one, even when everything else is washed-out and lackluster, the water carries its own lights. I set out for the railroad tracks, for the hill the flocks fly over, for the woods where the white mare lives. But first I go to the water.

Today is one of those excellent January partly cloudies in which light chooses an unexpected part of the landscape to trick out in gilt, and then shadow sweeps it away. You know you're alive. You take huge steps, trying to feel the planet's roundness arc between your feet. Kazantzakis says that when he was young he had a canary and a globe. When he freed the canary, it would perch on the globe and sing. All his life, wandering the earth, he felt as though he had a canary on top of his head, singing.

West of the house, Tinker Creek makes a sharp loop,

so that the creek is both in back of the house, south of me, and also on the other side of the road, north of me. I like to go north. There the afternoon sun hits the creek just right, deepening the reflected blue and lighting the sides of trees on the banks. Steers from the pasture across the creek come down to drink; I always flush a rabbit or two there. I sit on a fallen trunk in the shade and watch the squirrels in the sun. There are two separated wooden fences suspended from cables that cross the creek just upstream from my tree-trunk bench. These fences keep the steers from escaping up or down the creek when they come to drink. Squirrels, the neighborhood children, and I use the downstream fence as a swaying bridge across the creek. But the steers are there today.

I sit on the downed tree and watch the black steers slip on the creek bottom. They are all bred beef: beef heart, beef hide, beef hocks. They're a human product, like rayon. They're like a field of shoes. They have cast-iron shanks and tongues like foam insoles. You can't see through to their brains as you can with other animals; for there is beef fat behind their eyes, beef stew.

I cross the fence six feet above the water, walking my hands down the rusty cable and tightroping my feet along the narrow edge of the planks. When I hit the other bank and terra firma, some steers are bunched in a knot between me and the barbed-wire fence I want to cross. So I suddenly rush at them in a wild sprint, flailing my arms and hollering, "Lightning! Copperhead! Swedish meatballs!" They flee, still in a knot, stumbling across the flat pasture. I stand with the wind on my face.

When I slide under the barbed-wire fence, cross a field, and run over a sycamore trunk felled across the water, I'm on a little island shaped like a tear in the middle of Tinker Creek. On one side of the creek is a steep forested bank; the water is swift and deep on that side of the island. On the other side is the level field I walked through next to the steers' pasture; the water between the field and the island is shallow and sluggish. In summer's low water, flags and bulrushes grow along a series of shallow pools cooled by the lazy current. Water striders patrol the surface film, crayfish hump along the silt bottom eating filth, frogs shout and glare, and shiners and small bream hide among roots from the sulky green heron's eye. I come to this island every month of the year. I walk around it, stopping and staring, or I straddle the sycamore log over the creek, curling my legs out of the water in winter, trying to read. Today I sit on dry grass at the end of the island by the slower side of the creek. I'm drawn to this spot. I come to it as to an oracle; I return to it as a man years later will seek out the battlefield where he lost a leg or an arm.

A couple of summers ago I was walking along the edge of the island to see what I could see in the water, and mainly to scare frogs. Frogs have an inelegant way of taking off from invisible positions on the bank just ahead of your feet, in dire panic, emitting a froggy "Yike!" and splashing into the water. Incredibly, this amused me, and, incredibly, it amuses me still. As I walked along the grassy edge of the island, I got better and better at seeing frogs in and out of the water. I learned to recognize, slowing down, the difference in texture of the light reflected from mudbank, water, grass, or frog. Frogs were flying all around me. At the end of the island I noticed a small green frog. He was exactly half in and half out of the water, looking like a schematic diagram of an amphibian. He didn't jump.

I crept closer. At last I knelt on the island's dead grass, lost, dumbstruck, staring at the frog in the creek not four feet away. He was a very small frog, with wide, dull eyes. And just as I looked at him, he slowly crumpled and began to sag. The spirit vanished from his eyes as if snuffed. His skin emptied and drooped; his skull itself seemed to collapse and settle like a kicked tent. He was shrinking before my eyes like a deflating football. I watched the taut, glistening skin on his shoulders ruck, and rumple, and fall. Soon, part of his skin, formless as a pricked balloon, lay in floating folds like bright scum on top of the water: It was a monstrous and terrifying thing. I gaped bewildered, appalled. An oval shadow hung in the water just behind the drained frog; then the shadow glided away. The frog skin bag started to sink.

I had read about the giant water bug, but never seen one. "Giant water bug" is in fact the name of the creature, which is an enormous, heavy-bodied brown bug. It eats insects, tadpoles, fish, and frogs. Its grasping forelegs are mighty and hooked inward. It seizes a victim with these legs, hugs it tight, and paralyzes it with enzymes injected during a vicious bite. That one bite is the only bite it ever takes. Through the puncture shoot the poisons that dissolve the victim's muscles and bones and organs—all but the skin. Then the giant water bug sucks out the victim's body, reduced to a juice. This event is quite common in warm freshwater. And now I'd seen it myself. I was still kneeling on the island grass when the unrecognizable flag of frog skin settled on the creek bottom, swaying. I stood up and brushed the knees of my pants. I couldn't catch my breath.

Many carnivorous animals, of course, devour their prey alive. The usual method seems to be to subdue the victim by downing or grasping it so it can't flee, then eating it whole or in a series of bloody bites. Frogs eat everything whole, stuffing prey in their mouth with their thumbs. People have seen frogs with their wide jaws so full of live dragonflies they couldn't close them. Ants don't even have to catch their prey: In the spring they swarm over newly hatched, featherless birds in the nest and eat them, tiny bite by tiny bite.

That it's rough out there and chancy is no surprise. Every live thing is a survivor on a kind of extended emergency bivouac. But at the same time we are also created. In the Qur'an, Allah asks, "The heaven and the earth and all in between, thinkest thou I made them *in jest*?" It's a good question. What do we think of the created universe, spanning an unthinkable void with an unthinkable profusion of forms? And what do we think of nothingness, those sickening reaches of time in either direction?

If the giant water bug was not made in jest, was it then made in earnest? Pascal uses a nice term to describe the notion of the creator's once having called forth the universe, turning his back to it: *Deus absconditus*. Is this what we think happened? Was the sense of it there, and God absconded with it, ate it, like a wolf who disappears round the edge of the house with the Thanksgiving turkey? "God is subtle," Einstein said, "but not malicious." Einstein also said that "nature conceals her mystery by means of her essential grandeur, not by her cunning."

It could be that God has not absconded but spread, as our vision and understanding of the universe have spread, to a fabric of spirit and sense so grand and subtle, so powerful in a new way, that we can only feel blindly of its hem. In making the thick darkness a swaddling band for the sea, God "set bars and doors" and said, "Hitherto shalt thou come, but no further." But have we come even that far? Have we rowed out to the thick darkness, or are we all playing pinochle in the bottom of the boat?

Cruelty is a mystery, and the waste of pain. But if we describe a world to encompass these things, a world that is a long, brute game, then we bump against another mystery: the inrush of power and light, the canary that sings on the skull. For unless all ages and races of men have been deluded by the same mass hypnotist (who?), there seems to be such a thing as beauty, a grace wholly gratuitous.

About five years ago I saw a mockingbird make a straight vertical descent from the roof gutter of a four-story building. It was an act as careless and spontaneous as the curl of a stem or the kindling of a star. The mockingbird took a single step into the air and dropped. His wings were still folded against his sides as though he were singing from a limb and not falling, accelerating thirty-two feet per second per second, through empty air. Just a breath before he would have been dashed to the ground, he unfurled his wings with exact, deliberate care, revealing the broad bars of white, spread his elegant, white-banded tail, and so floated onto the grass. I had just rounded a corner when his insouciant step off the gutter caught my eye; there was no one else in sight. The fact of his free fall was like the old philosophical conundrum about the tree that falls in the forest. The answer must be, I think, that beauty and grace are performed whether or not we will or sense them. The least we can do is try to be there.

Another time I saw a different wonder: sharks off the Atlantic coast of Florida. There is a way a wave rises above the ocean horizon, a triangular wedge against the sky. If you stand where the ocean breaks on a shallow beach, you see the raised water in a wave is in fact translucent, shot with lights. One late afternoon at low tide a hundred big sharks passed the beach near the mouth of a tidal river in a feeding frenzy. As each green wave rose from the churning water, it illuminated within itself the six- or eight-foot-long bodies of twisting sharks. The sharks disappeared as each wave rolled toward me; then a new wave swelled above the horizon, containing in it, like scorpions in amber, sharks that roiled and heaved.

The sight held power and beauty, grace tangled in a rapture with violence.

We don't know what's going on here. If these tremendous events are random combinations of matter run amok, the yield of millions of monkeys at millions of typewriters, then what is it in us, hammered out of those same typewriters, that they ignite? We don't know. Our life is a faint tracing on the surface of mystery, like the idle, curved tunnels of leaf miners on the face of a leaf. We must somehow take a wider view, look at the whole landscape, really see it, and try to describe what's going on here. Then we can at least wail the right question into the swaddling band of darkness, or, if it comes to that, choir the proper praise.

At the time of Lewis and Clark, setting the prairies on fire was a well-known signal that meant "Come down to the water." It was an extravagant gesture, but we can't do less. If the landscape reveals one certainty, it is that the extravagant gesture is the very start and stuff of creation. The universe has continued to deal in extravagances, flinging intricacies and colossi down eons of emptiness, heaping profusions on profligacies with ever-fresh vigor. The whole show has been on fire from the word go. I come down to the water to cool my eyes. But everywhere I look I see fire; that which isn't flint is tinder and the whole world sparks and flames.

I have come to the grassy island late in the day. The creek is up; icy water sweeps under the sycamore log bridge. The frog skin, of course, is utterly gone. I have stared at that one spot on the creek bottom for so long, focusing past the rush of water, that when I stand, the opposite bank seems to stretch before my eyes and flow grassily upstream. When the bank settles down I cross the sycamore log and enter again the big plowed field next to the steers' pasture.

The wind is terrific out of the west; the sun comes and goes. I can see the shadow on the field before me deepen uniformly and spread like a plague. Everything seems so dull now I am amazed I can even distinguish objects. And suddenly the light runs across the land like a great comber, and up the trees, only to go again in a wink: I think I've gone blind or died. When it comes again, the light, you hold your breath, and if it stays you forget about it until it goes again.

It's the most beautiful day of the year. At four o'clock the eastern sky is a dead stratus black flecked with low white clouds. The sun in the west illuminates the ground, the mountains, and especially the bare branches of trees, so that everywhere silver trees cut into the black sky like a photographer's negative of a landscape. The air and the ground are dry; the mountains are flashing on and off like neon signs. Clouds slide east if pulled from the horizon, like a tablecloth whipped off a table. The hemlocks by the barbed-wire fence are flinging themselves east as though their backs would break. Purple shadows are racing east; the wind makes me face east, and again I feel dizzied and drawn, as I felt when the creek bank reeled.

At four-thirty the sky in the east is clear; how could that big blackness be blown? Fifteen minutes later another darkness comes overhead from the northwest; and it's here to stay. Everything drains of its light as if sucked. Only at the horizon do inky black mountains give way to distant, lighted mountains—lighted not by direct illumination but rather by paled glowing sets of mist hung before them. Now the blackness is in the east; everything is half in shadow, half in sun, every cloud, tree, mountain, and hedge. I can't see Tinker Mountain through the line of hemlock, till it comes on like a streetlight, ping, *ex nihilo*. Its sandstone cliffs pink and swell. Suddenly the light goes; the cliffs recede as if pushed. The sun hits a clump of sycamores between me and the mountains; the sycamore arms light up, and I can't see the cliffs. They're gone. The pale network of sycamore arms, which a second ago was transparent as a screen, is suddenly opaque, glowing with light. Now the sycamore arms snuff out, the mountains come on, and there are the cliffs again.

I walk home. By five-thirty the show has pulled out. Nothing is left but an unreal blue and a few banked clouds low in the north. Some sort of carnival magician has been here, some fast-talking worker of wonders who has the act backwards. "Something in this hand," he says, "something in this hand, something up my sleeve, something behind my back . . ." and abracadabra, he snaps his fingers: and it's all gone. Only the bland, blank-faced magician remains, in his unruffled cot, bare-handed, nodding at a smattering of baffled applause. When you look again the whole show has pulled up stakes and moved on down the road. It never stops. New shows roll in from over the mountains and the magician reappears unannounced from a fold in the curtain you never dreamed was an opening. Scarves of clouds, rabbits in plain view, disappear into the black hat forever. Presto chango. The audience, if there is one, is dizzy from head-turning, dazed.

Like the bear who went over the mountain, I went out to see what I could see. And, I might as well warn you, like the bear, all that I could see was the other side of the mountain: more of same. On a good day I might catch a glimpse of another wooded ridge rolling under the sun like water, another bivouac. I propose to keep here what Thoreau called "a meteorological journal of the mind," telling some tales and describing some of the sights of this rather tamed valley, and exploring, in fear and trembling, some of the unmapped dim reaches and unholy fastnesses to which those tales and sights so dizzyingly lead.

I am no scientist. I explore the neighborhood. An infant who has just learned to hold his head up has a frank and forthright way of gazing about him in bewilderment. He hasn't the faintest clue where he is, and he aims to learn. In a couple of years, what he will have learned instead is how to fake it: He'll have the cocksure air of a squatter who has come to feel he owns the place. Some unwonted, taught pride diverts us from our original intent, which is to explore the neighborhood, view the landscape, to discover at least *where* it is that we have been so startlingly set down, if we can't learn why.

So I think about the valley. It is my leisure as well as my work, a game. It is a fierce game I have joined because it is being played anyway, a game of both skill and chance, played against an unseen adversary—the conditions of time—in which the payoff, which may arrive at any moment in a blast of light, might as well come to me as anyone else. I stake the time I'm grateful to have, the energies I'm glad to direct. I risk getting stuck on the board, so to speak, unable to move in any direction, which happens enough, God knows; and I risk the searing, exhausting nightmares that plunder rest and force me facedown all night long in some muddy ditch seething with hatching insects and crustaceans.

But if I can bear the nights, the days are a pleasure. I walk out; I see something, some event I'd otherwise have utterly missed and lost; or something sees me, some enormous power brushes me with its clean wing, and I resound like a beaten bell.

I am an explorer, then, and I am also a stalker, or the instrument of the hunt itself. Certain Indians used to carve long grooves along the wooden shafts of the arrows. They called the grooves "lightning marks," because they resembled the curved fissure lightning slices down the trunks of trees. The function of lightning marks is this: If the arrow fails to kill the game, blood from a deep wound will channel along the lightning mark, streak down the arrow shaft, and spatter to the ground, laying a trail dripped on broad leaves, on stones, that the barefoot and trembling archer can follow into whatever deep or rare wilderness it leads. I am the arrow shaft, carved along my length by unexpected lights and gashes from the very sky, and this book is the straying trail of blood.

Something pummels us, something barely sheathed. Power broods and lights. We're played on like a pipe; our breath is not our own. James Houston describes two young Inuit girls sitting cross-legged on the ground, mouth on mouth, blowing by turns each other's throat cords, making a low, unearthly music. When I cross again the bridge that is really the steers' fence, the wind has thinned to the delicate air of twilight; it barely ruffles the water's skin. I watch the running sheets of light raised on the creek's surface. The sight has the appeal of the purely passive, like the racing of light under clouds on a field, the beautiful dream at the moment of being dreamed. The breeze is the merest puff, but you yourself sail headlong and breathless under the gale force of the spirit.

SEEING

WHEN I WAS SIX OR SEVEN YEARS OLD, growing up in Pittsburgh, I used to take a precious penny of my own and hide it for someone else to find. It was a curious compulsion; sadly, I've never been seized by it since. For some reason I always "hid" the penny along the same stretch of sidewalk up the street. I would cradle it at the roots of a sycamore, say, or in a hole left by a chipped-off piece of sidewalk. Then I would take a piece of chalk, and, starting at either end of the block, draw huge arrows leading up to the penny from both directions. After I learned to write I labeled the arrows: SURPRISE AHEAD or MONEY THIS WAY. I was greatly excited, during all this arrow drawing, at the thought of the first lucky passer-by who would receive in this way, regardless of merit, a free gift from the universe. But I never lurked about. I would go straight home and not give the matter another thought, until, some months later, when I'd be gripped again by the impulse to hide another penny.

It is still the first week in January, and I've got great plans. I've been thinking about seeing. There are lots of things to see, unwrapped gifts and free surprises. The world is fairly studded and strewn with pennies cast broadside from some generous hand. But—and this is the point—who gets excited by a mere penny? If you follow one arrow, if you crouch motionless on a bank to watch a tremendous ripple thrill on the water and find yourself rewarded by the sight of a muskrat kit paddling from its den, will you count that sight a chip of copper only, and go your rueful way?

It is dire poverty indeed when a man is so hungry and tired that he won't stoop to pick up a penny. But if you cultivate a healthy poverty and simplicity, so that finding a penny will literally make your day, then, since the world is in fact planted in pennies, you have with your poverty bought a lifetime of days. It is that simple. What you see is what you get.

I used to be able to see flying insects in the air. I'd look ahead and see, not the row of hemlocks across the road, but the air in front of it. My eyes focused along that column of air, picking out flying insects. But I lost interest, I guess, for I dropped the habit. Now I can see birds. Probably some people can look at the grass at their feet and discover all the crawling creatures. I would like to know grasses and sedges—and care. Then my least journey into the world would be a field trip, a series of happy recognitions. Thoreau, in an expansive mood, exalted, "What a rich book might be made about buds, including, perhaps, sprouts!" It would be nice to think so. I cherish mental images I have of three perfectly happy people. One collects stones. Another—an Englishman, say—watches clouds. The third lives on a coast and collects drops of seawater which he examines microscopically and mounts. But I don't see what the specialist sees, and so I cut myself off, not only from the total picture but from the various forms of happiness.

Unfortunately, nature is very much a now-you-see-it, now-you-don't affair. A fish flashes, then dissolves in the water before my eyes like so much salt. Deer apparently ascend bodily into heaven; the brightest oriole fades into leaves. These disappearances stun me into stillness and concentration. They say of nature that it conceals with a grand nonchalance, and they say of vision that it is a deliberate gift, the revelation of a dancer who for my eyes only flings away her seven veils. For nature does reveal as well as conceal: now-you-don't-see-it, now-you-*do*. For a week last September, migrating red-winged blackbirds were feeding heavily down by the creek at the back of the house. One day I went out to investigate the racket; I walked up to a tree, an Osage orange, and a hundred birds flew away. They simply materialized out of the tree. I saw a tree, then a whisk of color, and then a tree again. I walked closer still and a hundred other blackbirds took flight. Not a branch, not a twig budged: The birds were as weightless, apparently, as they were invisible. It was as if the leaves of the Osage orange had been freed from a spell in the form of red-winged blackbirds; they flew from the tree, caught my eye in the sky, and vanished. When I looked again at the tree, its leaves had reassembled as if nothing had happened. Finally I walked directly to the trunk of the tree and a final hundred, the real diehards, appeared, spread, and vanished. How could so many birds hide in the tree without my seeing them? The Osage orange, unruffled, looked just as it had looked from the house, when three hundred red-winged blackbirds cried from its crown. I looked downstream to where they flew, but they were gone; I couldn't spot a single one. I wandered downstream to force them to play their hand, but they'd crossed the creek and scattered. One show to a customer. These appearances catch at my throat; they are the free gifts, the bright coppers at the roots of trees.

It's all a matter of keeping my eyes open. Nature is like one of those line drawings of a tree that are puzzles for children: Can you find hidden in the leaves a duck, a house, a boy, a bucket, a zebra, and a boot? Specialists can find the most incredibly well-hidden things. A book I read when I was young recommended an easy way to find caterpillars to rear: You simply find some fresh caterpillar droppings, look up, and there's your caterpillar. More recently an author advised me to set my mind at ease about those piles of cut stems on the ground in grassy fields. Field mice make them; they cut the grass down by degrees to reach the seeds at the head. It seems that when the grass is tightly packed, as in a field of ripe grain, the blade won't topple at a single cut through the stem; instead, the cut stem simply drops vertically, held in the crush of grain. The mouse severs the bottom again and again, the stem keeps dropping an inch at a time, and finally the head is low enough for the mouse to reach the seeds. Meanwhile, the mouse is positively littering the field with its little piles of cut stems into which, presumably, the author of that book is constantly stumbling.

If I can't make out these minutiae, still I try to keep my eyes open. I'm always on the lookout for ant lion traps in sandy soil, monarch pupae near milkweed, skipper larvae in locust leaves. These things are utterly common, and I've not seen one. I bang on hollow trees near water, but so far no flying squirrels have appeared.

In flat country I watch every sunset in hopes of seeing the green flash—a seldom-seen streak of light that rises from the sun, like a spurting fountain, at the moment of sunset; it throbs into the sky for two seconds and disappears. One more reason to keep my eyes open. A photography professor at the University of Florida just happened to see a bird die in midflight; it jerked, died, dropped, and smashed on the ground. I squint at the wind because I read Stewart Edward White: "I have always maintained that if you looked closely enough you could see the wind—the dim, hardly-made-out, fine débris fleeing high in the air." White was an excellent observer, and devoted an entire chapter of *The Mountains* to the subject of seeing deer: "As soon as you can forget the naturally obvious and construct an artificial obvious, then you too will see deer."

But the artificial obvious is hard to see. My eyes account for less than 1 percent of the weight of my head; I'm bony and dense; I see what I expect. I once spent a full three minutes looking at a bullfrog that was so unexpectedly large I couldn't see it, even though a dozen enthusiastic campers were shouting directions. I finally asked, "What color am I looking for?" and a fellow said, "Green." When at last I picked out the frog, I saw what painters are up against: The thing wasn't green at all, but the color of wet tree bark.

The lover can see, and the knowledgeable. I visited an aunt and uncle at a quarter-horse ranch in Cody, Wyoming. I couldn't do much of anything useful, but I could, I thought, draw. So, as we all sat around the kitchen table after supper, I produced a sheet of paper and drew a horse. "That's one lame horse," my aunt volunteered. The rest of the family joined in: "Only place to saddle that one is his neck"; "Looks like we better shoot the poor thing, on account of those terrible growths." Meekly I slid the pencil and paper down the table. Everyone in that family, including my three young cousins, could draw a horse. Beautifully. When the paper came back it looked as though five shining, real quarter horses had been corralled by mistake with a papier-mâché moose; the real horses seemed to gaze at the monster with a steady, puzzled air. I stay away from horses now, but I can do a credible goldfish. The point is that I just don't know what the lover knows; I just can't see the artificial obvious that those in the know construct. The herpetologist asks the native, "Are there snakes in that ravine?" "Nosir." And the herpetologist comes home with, yessir, three bags full. Are there butterflies on that mountain? Are the bluets in bloom, are there arrowheads here, or fossil shells in the shale?

Peeping through my keyhole I see within the range of only about 30 percent of the light that comes from the sun; the rest is infrared and some little ultraviolet, perfectly apparent to many animals, but invisible to me. As for what I do see, a nightmare network of ganglia, charged and firing without my knowledge, cuts and splices it, editing it for my brain. Donald E. Carr points out that the sense impressions of one-celled animals are not edited for the brain: "This is philosophically interesting in a rather mournful way, since it means that only the simplest animals perceive the universe as it is."

A fog that won't burn away drifts and flows across my field of vision. I can't distinguish the fog from the overcast sky; I can't be sure if the light is direct or reflected. When you see fog move against a backdrop of deep pines, it's not the fog itself you see, but streaks of clearness floating across the air in shreds. So I see only tatters of clearness through a pervading obscurity.

Everywhere darkness and the presence of the unseen appalls. We estimate now that only one atom dances alone in every cubic meter of intergalactic space. I blink and squint. What planet or power yanks Halley's Comet out of orbit? We haven't seen that force yet; it's a question of distance, density, and the pallor of reflected light. We rock, cradled in the swaddling band of darkness. Even the simple darkness of night whispers suggestions to the mind. Last summer, in August, I stayed at the creek too late.

Where Tinker Creek flows under the sycamore log bridge to the tear-shaped island, it is slow and shallow, fringed thinly in cattail marsh. Here, vast breeding populations of insects, fish, reptiles, birds, and mammals collect. On windless summer evenings I stalk along the creek bank or straddle the sycamore log in absolute stillness, watching for muskrats. The night I stayed too late I was hunched on the log staring spellbound at spreading stains of lilac reflected on the water. A cloud in the sky suddenly lighted as if turned on by a switch. Its reflection just as suddenly materialized on the water upstream, flat and floating, so that I could no longer see the creek bottom. Downstream, away from the cloud on the water, water turtles smooth as beans were gliding down with the current in a series of easy, weightless push-offs, as men bound on the moon. I didn't know whether to trace the progress of one turtle I was sure of, risking sticking my face in one of the bridge's spiderwebs made invisible by the gathering dark, or to take a chance on seeing the carp, or to scan the mudbank in hope of seeing a muskrat, or to follow the last of the swallows who caught at my heart and trailed it after them like streamers as they appeared from directly below, under the log, flying upstream with their tails forked, fast.

But shadows spread, and deepened, and stayed. After thousands of years, we're still strangers to darkness, fearful aliens in an enemy camp with our arms crossed over our chests. I stirred. A land turtle on the bank, startled, hissed the air from its lungs and withdrew into its shell. An uneasy pink here, an unfathomable blue there, gave great suggestion of lurking beings. Things were going on. I heard a dry rattle; a not-so-distant rattlesnake, slit-eyed and sinuous, or just a towhee kicking the leaves? Action roiled the water everywhere I looked, big action, inexplicable. A tremor welled up beside a gaping musk-rat burrow in the bank and I caught my breath, but no muskrat appeared. The ripples continued to fan up-stream. Night was knitting over my face an eyeless mask. Still I sat transfixed. A distant airplane, a delta wing out of nightmare, made a gliding shadow on the creek's bottom like a stingray cruising upstream. At once a black fin slit the pink cloud on the water, shearing it in two. These two halves merged and then dissolved before my eyes. Darkness pooled in the cleft of the creek and rose, as water collects in a well. Untamed, dreaming lights flickered over the sky. I saw hints of hulking underwater shadows, two pale splashes out of the water, and round ripples rolling close together from a blackened center.

At last I stared upstream where only the deepest violet remained of the cloud, a cloud so high its underbelly still glowed feeble color reflected from a hidden sky lighted in turn by a sun halfway to China. And out of violet, a sudden enormous black body arced over the water. I saw only cylindrical sleekness. Head and tail, if there was a head and tail, were both submerged in cloud. I saw only one ebony fling, a headlong dive to darkness. Then the waters closed, and the lights went out.

I walked home in a shivering daze, uphill and down. Later I lay openmouthed in bed, my arms flung wide at my sides to steady the whirling darkness. At this latitude I'm spinning 836 miles an hour round the earth's axis; I often fancy I feel my sweeping fall as a breakneck arc like the dive of dolphins, and the hollow rushing of wind raises hair on my neck and the side of my face. In orbit around the sun I'm moving 64,800 miles an hour. The solar system as a whole, like a merry-go-round unhinged, spins, bobs, and blinks at the speed of 43,200 miles an hour along a course set east of Hercules. Someone has piped, and we are dancing a tarantella until the sweat pours. I open my eyes and I see dark, muscled forms curl out of water, with flapping gills and flattened eyes. I close my eyes and I see stars, deep stars giving way to deeper stars, deeper stars bowing to deepest stars at the crown of an infinite cone.

"Still," wrote Van Gogh in a letter, "a great deal of light falls on everything." If we are blinded by darkness, we are also blinded by light. When too much light falls on everything, a special terror results. Peter Freuchen describes the notorious kayak sickness to which Greenland Inuits are prone. "The Greenland fjords are peculiar for the spells of completely quiet weather, when there is not enough wind to blow out a match and the water is like a sheet of glass. The kayak hunter must sit in his boat without stirring a finger so as to not scare the shy seals away. . . . The sun, low in the sky, sends a glare into his eyes, and the landscape around moves into the realm of the unreal. The reflex from the mirrorlike water hypnotizes him, he seems to be unable to move, and all of a sudden it is as if he were floating in a bottomless void, sinking, sinking, and sinking. . . . Horror-stricken, he tries to stir, to cry out, but he cannot, he is completely paralyzed, he just falls and falls." Some hunters are especially cursed with this panic, bringing ruin, even starvation, to their families.

Sometimes here in Virginia at sunset low clouds on the southern or northern horizon are completely invisible in the lighted sky. I only know one is there because I can see its reflection in still water. The first time I discovered this mystery I looked from cloud to no-cloud in bewilderment, checking my bearings over and over, thinking maybe the ark of the covenant was just passing by south of Dead Man Mountain. Only much later did I read the explanation: Polarized light from the sky is very much weakened by reflection, but the light in clouds isn't polarized. So invisible clouds pass among visible clouds, till all slide over the mountains; so a greater light extinguishes a lesser as though it didn't exist.

In the great meteor shower of August, the Perseid, I wail all day for the shooting stars I miss. They're out there showering down, committing hara-kiri in a flame of fatal attraction, and hissing perhaps at last into the ocean. But at dawn what looks like a blue dome clamps down over me like a lid on a pot. The stars and planets could smash and I'd never know. Only a piece of ashen moon occasionally climbs up or down the inside of the dome, as our local star without surcease explodes over our heads. We have really only that one light, one source for all power, and yet we must turn away from it by universal decree. Nobody here on the planet seems aware of this strange, powerful taboo, that we all walk about carefully averting our faces, this way and that, lest our eyes be blasted forever.

Darkness appalls and light dazzles; the scrap of visible light that doesn't hurt my eyes hurts my brain. What I see sets me swaying. Size and distance and the sudden swelling of meanings confuse me, bowl me over. I straddle the sycamore log bridge over Tinker Creek in summer. I look at the lighted creek bottom: Snail tracks tunnel the mud in quavering curves. A crayfish jerks, but by the time I absorb what has happened, he's gone in a billowing smoke screen of silt. I look at the water: minnows and shiners. If I'm thinking minnows, a carp will fill my brain till I scream. I look at the water's surface: skaters, bubbles, and leaves sliding down. Suddenly, my own face, reflected, startles me witless. Those snails have been tracking my face! Finally, with a wrench of the will, I see clouds, cirrus clouds. I'm dizzy, I fall in. This looking business is risky.

Once I stood on a humped rock on nearby Purgatory Mountain, watching through binoculars the great autumn hawk migration below, until I discovered that I was in danger of joining the hawks on a vertical migration of my own. I was used to binoculars, but not, apparently, to balancing on humped rocks while looking through them. I staggered. Everything advanced and receded by turns; the world was full of unexplained foreshortenings and depths. A distant huge tan object, a hawk the size of an elephant, turned out to be the browned bough of a nearby loblolly pine. I followed a sharp-shinned hawk against a featureless sky, rotating my head unawares as it flew, and when I lowered the glass a glimpse of my own looming shoulder sent me staggering. What prevents the men on Palomar from falling, voiceless and blinded, from their tiny, vaulted chairs?

I reel in confusion; I don't understand what I see. With the naked eye I can see two million light-years to the Andromeda galaxy. Often I slop some creek water in a jar, and when I get home, I dump it in a white china bowl. After the silt settles I return and see the tracings of minute snails on the bottom, a planarian or two winding round the rim of water, roundworms shimmying frantically, and finally, when my eyes have adjusted to these dimensions, amoebae. At first the amoebae look like muscae volitantes, those curled moving spots you seem to see in your eyes when you stare at a distant wall. Then I see the amoebae as drops of water congealed, bluish, translucent, like chips of sky in the bowl. At length I choose one individual and give myself over to its idea of an evening. I see it dribble a grainy foot before it on its wet, unfathomable way. Do its unedited sense impressions include the fierce focus of my eyes?

Shall I take it outside and show it Andromeda, and blow its little endoplasm? I stir the water with a finger, in case it's running out of oxygen. Maybe I should get a tropical aquarium with motorized bubbles and lights, and keep this one for a pet. Yes, it would tell its fissioned descendants, the universe is two feet by five, and if you listen closely you can hear the buzzing music of the spheres.

Oh, it's mysterious lamplit evenings, here in the galaxy, one after the other. On one of those nights, I wander from window to window, looking for a sign. But I can't see. Terror and a beauty insoluble are a ribbon of blue woven into the fringes of garments of things both great and small. No culture explains, no bivouac offers real heaven or rest. But it could be that we are not seeing something. Galileo thought comets were just an optical illusion. This is fertile ground: Since we are certain that they're not, perhaps we can look at what our scientists have been saying with fresh hope. What if there are really gleaming, castellated cities hung upside down over the desert sands? What limpid lakes and cool date palms may our caravans have passed untried? Until, one by one, by the blindest of leaps, we light on the road to these places, we must stumble in darkness and hunger.

I turn from the window. I'm blind as a bat, sensing from every direction only the echo of my own thin cries.

Seeing is of course very much a matter of verbalization. Unless I call my attention to what passes before my eyes, I simply won't see it. It is, as Ruskin says, "not merely unnoticed, but in the full, clear sense of the word, unseen." My eyes alone can't solve analogy tests using figures, the ones that show, with increasing elaborations, a big square, then a small square in a big square, then a bigger triangle. I have to say the words, describe what I'm seeing. If Tinker Mountain erupted, I'd likely notice. But if I want to notice the lesser cataclysms of valley life, I must maintain in my head a running description of the present. It's not that I'm especially observant; it's just that I talk too much. Otherwise, especially in a strange place, I'll never know what's happening. Like a blind man at the ball game, I need a radio.

When I see this way I analyze and pry. I hurl over logs and roll away stones; I study the bank a square foot at a time, probing and tilting my head. Some days when a mist covers the mountains, when the muskrats won't show and the microscope's mirror shatters, I want to climb up the blank blue dome as a man would storm the inside of a circus tent, wildly, dangling, and, with a steel knife, claw a rent in the top to peep out, even at the risk of a fall.

But there is another kind of seeing that involves a letting go. When I see this way I sway, transfixed and emptied. The difference between the two ways of seeing is the difference between walking with and without a camera. When I walk with a camera I walk from shot to shot, reading the light on a calibrated meter. When I walk without a camera, my own shutter opens, and the moment's light prints on my own silver gut. When I see this second way I am above all an unscrupulous observer.

It was sunny one evening last summer at Tinker Creek; the sun was low in the sky, upstream. I was sitting on the sycamore log bridge with the sunset at my back, watching shiners the size of minnows feeding over the muddy sand in skittery schools. Again and again, one fish, then another, turned for a split second across the current and—flash!—the sun shot out from its silver side. I couldn't watch for it. It was always just happening somewhere else, and it drew my vision just as it disappeared: flash, like a sudden dazzle of the thinnest blade, a sparking over a dun and olive ground at chance intervals from every direction. Then I noticed white specks, some sort of pale petals, small, floating from under my feet on the creek's surface, very slow and steady. So I blurred my eyes and gazed toward the brim of my hat and saw a new world. I saw the pale white circles roll up, roll up, like the world's turning, mute and perfect, and I saw the linear flashes, gleaming silver, like stars being born at random down a rolling scroll of time. Something broke and something opened. I filled up like a new wineskin. I breathed an air like light; I saw a light like water. I was the lip of a fountain the creek filled forever; I was ether, the leaf in zephyr; I was a flesh-flake, feather, bone.

When I see this way I see truly. As Thoreau says, I return to my senses. I am the man who watched the baseball game in silence in an empty stadium. I see the game purely, abstracted and dazed. When it's all over and the white-suited players lope off the green field to their shadowed dugouts, I leap to my feet and cheer.

But I can't go out and try to see this way. I'll fail, I'll go mad. All I can do is try to gag the commentator, to hush the noise of useless interior babble that keeps me from seeing just as surely as a newspaper dangled before my eyes. The effort is really a discipline requiring a lifetime of dedicated struggle; it marks the literature of saints and monks of every order East and West, under every rule and no rule, discalced and shod. The world's spiritual geniuses seem to discover universally that the mind's muddy river, this ceaseless flow of trivia and trash, cannot be dammed, and that trying to dam it is a waste of effort that might lead to madness. Instead you must allow the muddy river to flow unheeded in the dim channels of consciousness; you raise your sights; you look along it, mildly, acknowledging its presence without interest and gazing beyond it into the realm of the real where subjects and objects act and rest purely, without utterance. "Launch into the deep," says Jacques Ellul, "and you shall see."

The secret of seeing is, then, the pearl of great price. If I thought he could teach me to find it and keep it forever I would stagger barefoot across a hundred deserts after any lunatic at all. But while the pearl may be found, it may not be sought. The literature of illumination reveals this above all: Although it comes to those who wait for it, it is always, even to the most practiced and adept, a gift and a total surprise. I return from one walk knowing where the killdeer nests in the field by the creek and the hour the laurel blooms. I return from the same walk a day later scarcely knowing my own name. Litanies hum in my ears; my tongue flaps in my mouth Ailinon, alleluia! I cannot cause light; the most I can do is try to put myself in the path of its beam. It is possible, in deep space, to sail on solar wind. Light, be it particle or wave, has force: You rig a giant sail and go. The secret of seeing is to sail on solar wind. Hone and spread your spirit till you yourself are a sail, whetted, translucent, broadside to the merest puff.

When European and American surgeons learned to remove cataracts, many blind people could suddenly see. (Those blind from birth, however, could make no sense of their visual fields, saying they saw only "color patches." They had never learned as babies to distinguish near objects from far ones.) Many refused to use this incomprehensible new sight of theirs. Others, who had known objects only by feel, began learning to see.

A girl led into a garden "stands speechless in front of the tree, which she names 'tree' only by taking hold of it, and then as 'the tree with the lights in it.' "

It was for this tree I searched through the peach orchards of summer, in the forests of fall and down winter and spring for years. Then one day, walking along Tinker Creek, thinking of nothing at all, I saw it—the tree with the lights in it. It was the same backyard cedar where the mourning doves roost, only charged and transfigured, each cell buzzing with flame. I stood on the grass with the lights in it, grass that was wholly fire, utterly focused and utterly dreamed. It was less like seeing than like being for the first time seen, knocked breathless by a powerful glance.

Gradually the lights went out in the cedar, the colors died, the cells unflamed and disappeared. I was still ringing. I had been my whole life a bell, and never knew it until at that moment I was lifted and struck.

THE WATERS OF SEPARATION

Fair weather cometh out of the north: with God is terrible majesty.

—THE BOOK OF JOB, 37:22

TODAY IS THE WINTER SOLSTICE. The planet tilts just so to its star, lists and holds circling in a fixed tension between veering and longing, spins helpless, exalted, in and out of that fleet blazing touch. Last night Orion vaulted and spread all over the sky, pagan and lunatic, his shoulder and knee on fire, his sword three suns at the ready—for what?

And today was fair, hot, even; I woke and my fingers were hot and dry to their own touch, like the skin of a stranger. I stood at the window, the bay window on which in summer a waxy-looking grasshopper had breathed puff puff, and thought, I won't see this year again, not again so innocent, and longing wrapped round my throat like a scarf. "For the Heavenly Father desires that we should see," said Ruysbreck, "and that is why He is ever saying to our inmost spirit one deep unfathomable word and nothing else." But what is that word? Is this mystery or coyness? A cast-iron bell hung from the arch of my rib cage; when I stirred it rang, or it tolled, a long syllable pulsing ripples up my lungs and down the gritty sap inside my bones, and I couldn't make it out; I felt the voiced vowel like a sigh or a note, but couldn't catch the consonant that might shape it into sense. I wrenched myself from the window and stepped outside.

Here by the mock-orange hedge was a bee, a honeybee, sprung from its hive by the odd heat. I had an idea. I had read recently that ancient Romans thought bees were killed by echoes. It seemed a far-fetched and pleasing nothing, that a spoken word or falling rock given back to its cliff—the airy nothing that nevertheless bears and spreads the uncomprehended impact of something—should stun these sturdy creatures right out of the air. I could put it to the test. It was as good an excuse for a walk as any; it might still the bell even, or temper it true.

I knew where to find an echo; I'd have to take my chances on finding another December bee. I tied a sweater around my waist and headed for the quarry. It was hot; I never needed the sweater. A great tall cloud moved elegantly across an invisible walkway in the upper air, sliding on its flat foot like some great, proud snail.

I smelled silt on the wind, turkey, laundry, leaves . . . my God, what a world. There is no accounting for one second of it. On the quarry path through the woods I saw a discarded aquarium. One side of its glass was shattered. I could plant a terrarium here, I thought; I could transfer the two square feet of forest floor *under* the glass to *above* the glass, framing it, hiding a penny, and saying to passersby: Look! Look! Here is two square feet of the world.

I waited for an hour at the quarry, roving, my eyes filtering the air for flecks, until at last I discovered a bee. It was flying listlessly among dried weeds on the stony bank where I had sat months ago and watched a mosquito pierce the neck of a copperhead and suck. Beyond the bank lay the quarry pond, partly iced still, backed by stone cliffs. The setup was perfect:

Hello!, I tried tentatively: Hello!, faltered the cliffs under the forest; and did the root tips quiver the rock? But that is no way to kill a creature, saying hello. Goodbye!, I shouted: Goodbye!, came back, and the bee drifted unconcerned among the weeds.

It could be, I reasoned, that ancient Roman naturalists knew this fact that has escaped us because it works only in Latin. My Latin is sketchy. *Habeas corpus!* I cried; *Deus absconditus! Veni!* And the rock cliff batted it back: *Veni!* And the bee droned on.

That was that. It was almost noon; the tall cloud was gone. I stood alone. I still seemed to hear the odd sound of my own voice honed to a quaver by rock, thrown back down my throat and cast dying around me, lorn. Could it have been heard at Hollins Pond? Was anybody there to hear? I felt again the bell resounding faint under my ribs. I'm coming, I thought, when I can. I quit the quarry, my spurt of exuberance drained, my spirit edgy and tight.

The quarry path parallels Tinker Creek far upstream from my house, and when the woods broke into clearing and pasture, I followed the creek banks down. When I drew near the tear-shaped island, which I had never before approached from this side of the creek, a fence barred my way, a wire fence that wobbled across the creek and served me as a sagging bridge to the island. I stood, panting, breathing the frail scent of fresh water, felt the sun heat my hair.

The December grass on the island was blanched and sere, pale against the dusty boles of sycamores, noisy underfoot. Behind me, the way I had come, rose the pasture belonging to Twilight, a horse of perpetually different color whose name was originally Midnight, and who one spring startled the neighborhood by becoming brown. Far before me Tinker Mountain glinted and pitched in the sunlight. The Lucas orchard spanned the middle distance, its wan peach limbs swept and poised just so, row upon row, like a stageful of thin, innocent dancers who will never be asked to perform. Below the orchard rolled the steers' pasture, the creek, and the island where I'd watched a green frog sucked to its skin. Vaulting overhead was a fugitive, empty sky.

Downstream at the island's tip, where the giant water bug had held the living frog, I sat and sucked at my own dry knuckles. It was the way the frog's eyes crumpled. His mouth was a gash of terror; the shining skin of his breast and shoulder shivered once and sagged, reduced to an empty purse. But, oh, those two snuffed eyes! They crinkled, the comprehension drained out of them, as if sense and life had been merely incidental to the idea of eyes—a filling like any jam in a jar that is easily emptied—and flattened, lightless and opaque. Did the giant water bug have the frog by the back parts, or by the hollow of the thigh? Would I eat a frog's leg if offered? Sure.

In Old Testament times, the priest sacrificed a ram as thanksgiving. Then he butchered it and waved the breast meat ("the wave breast") and heaved the shoulder meat ("the heave shoulder") at the altar of the Lord. What I want to know is this: Does the priest heave it *at* the Lord?

Does he *throw* the shoulder of the ram of consecration—a ram that, before the priest slayed and chunked it, had been perfect and whole, not "Blind, or broken, or maimed, or having a wen, or scurvy, or scabbed . . . bruised, or crushed, or broken, or cut"—does he hurl it across the tabernacle, between the blood-ied horns of the altar, hurl it at God? Now look at what you made me do. And then he eats it. This heave is a violent, desperate way of catching God's eye. It is not inappropriate. We are people; we are permitted to have dealings with the creator and we must speak up for the creation. God, *look* at what you've done to this creature, look at the sorrow, the cruelty, the long damned waste! Can it be possible, can it ludicrously be for *this* that on this unconscious planet with my innocent kind I play softball all spring, to develop my throwing arm?

How high, how far, could I heave a little shred of frog shoulder at the Lord? How high, how far, how long until I die?

I fingered the winter-killed grass, looping it round the tip of my finger like hair, ruffling its tips with my palms. Another year has twined away, unrolled and dropped across nowhere like a flung banner painted in gibberish. "The last act is bloody," said Pascal, "however brave be all the rest of the play; at the end they throw a little earth upon your head, and it's all over forever." Somewhere, everywhere, there is a gap, like the shuddering chasm of Shadow Creek which gapes open at my feet, like a sudden split in the window or hull of a high-altitude jet, into which things slip, or are blown, out of sight, vanished in a rush, blasted, gone, and can no more be found. For the living there is rending loss at each opening of the eye, each *Augenblick,* as a muskrat dives, a heron takes alarm, a leaf floats spinning away. There is death in the pot for the living's food, flyblown meat, muddy salt, and plucked herbs bitter as squill. If you can get it. How many people have prayed for their daily bread and famished? They die their daily death as utterly as did the frog, *people,* played with, dabbled upon, when God knows they loved their life. In a winter famine, desperate Algonquian Indians "ate broth made of smoke, snow, and buckskin, and the rash of pellagra appeared like tattooed flowers on their emaciated bodies—the roses of starvation," in a French physician's description, "and those who starved died covered with roses." Is this beauty, these gratuitous roses, or a mere display of force?

Or is beauty itself an intricately fashioned lure, the cruelest hoax of all? There is a certain fragment of an ancient and involved Inuit tale I read in Farley Mowat that for years has risen, unbidden, in my mind. The fragment is a short scenario observing all the classical unities, simple and cruel, and performed by the light of a soapstone seal-oil lamp.

A young man in a strange land falls in love with a young woman and takes her to wife in her mother's tent. By day the two women chew skins and boil meat while the young man hunts. But the old crone is jealous; she wants the boy. Calling her daughter to her one day, she offers to braid her hair; the girl sits pleased, proud, and is soon strangled by her own hair. One thing Inuit know is skinning. The mother takes her curved hand knife shaped like a dancing skirt, skins her daughter's beautiful face, and presses that empty flap smooth on her own skull. When the boy returns that night he lies with her, in that tent at the top of the world. He is wet from hunting, however, and soon the skin mask shrinks and slides, uncovering the shriveled face of the old mother. The boy flees in horror, forever.

Could it be that if I climbed the dome of heaven and scrabbled and clutched at the beautiful cloth till I loaded my fist with a wrinkle to pull, that the mask would rip away to reveal a toothless old crone, eyes glazed with delight?

A wind rose, quickening; it invaded my nostrils, vibrated my gut. I stirred and lifted my head. No, I've gone through this a million times, beauty is not a hoax—how many days have I learned not to stare at the back of my hand when I could look out at the creek? Come on, I say to the creek, surprise me; and it does, with each new drop. Beauty is real. I would never deny it; the appalling thing is that I forget it. Waste and extravagance go tougher up and down the banks, all along the intricate fringe of spirit's free incursions into time. On either side of me the creek snared and kept the sky's distant lights, shaped them into shifting substance and bore them speckled down.

This Tinker Creek! It was low today, and clear. On the still side of the island the water held pellucid as a pane, a gloss on runes of sandstone, shale, and silt. The fast side showed flecks of shadow and tatters of sky. There are the waters of beauty and mystery, issuing from a gap in the granite world; they fill the lodes in my cells with a light like petaled water, and they churn in my lungs mighty and frigid, like a big ship's screw. And these are also the waters of separation. They purify, acrid and laving, and they cut me off. I am spattered with a sop of ashes, burnt bone knobs, and blood; I range wild-eyed, flying over fields and plundering the woods, no longer quite fit for company.

In the old Hebrew ordinance for the waters of separation, the priest must find a red heifer, a red heifer unblemished, which has never known the yoke, and lead her outside the people's camp, and sacrifice her, burn her wholly, without looking away: "Burn the heifer in his sight; her skin, and her flesh, and her blood, with her dung, shall he burn." Into the stinking flame the priest casts the wood of a cedar tree for longevity, hyssop for purgation, and a scarlet thread for a vein of living blood. It is from these innocent ashes that the waters of separation are made, anew each time, by steeping them in a vessel with fresh running water. This special water purifies. A man—any man—dips a sprig of hyssop into the vessel and sprinkles—merely sprinkles!—the water upon the unclean, "upon him that touched a bone, or one slain, or one dead." So. But I never signed up for this role. The bone touched me.

I stood, alone, and the world swayed. I am a fugitive and a vagabond, a sojourner seeking signs. Isak Dinesen in Kenya, her heart utterly broken by loss, stepped out of the house at sunrise, seeking a sign. She saw a rooster lunge and rip a chameleon's tongue from its root in the throat and gobble it down. And then Isak Dinesen had to pick up a stone and smash the chameleon. But I had seen that sign—what she saw, that is, the world's cruelty—more times than I had ever sought it. Then today I saw an inspiriting thing, a pretty thing, really, and small.

I was standing lost, sunk, my hands in my pockets, gazing toward Tinker Mountain and feeling the earth reel down. All at once I saw what looked like a Martian spaceship whirling toward me in the air. It flashed borrowed light like a propeller. Its forward motion greatly outran its fall. As I watched, transfixed, it rose, just before it would have touched a thistle, and hovered pirouetting in one spot, then twirled on and finally came to rest. I found it in the grass; it was a maple key, a single winged seed from a pair. Hullo. I threw it into the wind and it flew off again, bristling with animate purpose—not like a thing dropped or windblown, pushed by the witless winds of convection currents hauling round the world's rondure where they must—but like a creature muscled and vigorous, or a creature spread thin to that other wind, the wind of the spirit which bloweth where it listeth, lighting, and raising up, and easing down. O maple key, I thought, I must confess I thought, oh welcome, cheers.

The bell under my ribs rang a true note, a flourish as of blended horns, clarion, sweet, and making a long dim sense I will try at length to explain. Flung is too harsh a word for the rush of the world. Blown is more like it, but blown by a generous, unending breath. That breath never ceases to kindle, exuberant, abandoned; frayed splinters spatter in every direction and burgeon into flame. And now when I sway to a fitful wind, alone and listing, I will think, maple key. When I see a photograph of earth from space, the planet so startlingly painterly and hung, I will think, maple key. When I shake your hand or meet your eyes I will think, two maple keys. And if I am a single maple key falling, at least I can twirl.

Thomas Merton wrote, "There is always a temptation to diddle around in the contemplative life, making itsy-bitsy statues." There is always an enormous temptation in all of life to diddle around making itsy-bitsy friends and meals and journeys for itsy-bitsy years on end. It is so self-conscious, so apparently moral, simply to step aside from the gaps where the creeks and winds pour down, saying, I never merited this grace, quite rightly, and then to sulk along the rest of your days on the edge of rage. I won't have it. The world is wilder than that in all directions, more dangerous and bitter, more extravagant and bright. We are making hay when we should be making whoopee; we are raising tomatoes when we should be raising Cain, or Lazarus.

Ezekiel excoriates false prophets as those who have "not gone up into the gaps." The gaps are the thing. The gaps are the spirit's one home, the altitudes and latitudes so dazzlingly spare and clean that the spirit can discover itself for the first time like a once-blind man unbound. The gaps are the cliffs in the rock where you cower to see the back parts of God; they are the fissures between mountains and cells the wind lances through, the icy narrowing fjords splitting the cliffs of mystery.

Go up into the gaps. If you can find them; they shift and vanish too. Stalk the gaps. Squeak into a gap in the solid, turn, and unlock—more than a maple—a universe. This is how you spend this afternoon, and tomorrow morning, and tomorrow afternoon. *Spend* the afternoon. You can't take it with you.

I live in tranquility and trembling. Sometimes I dream. I am interested in Alice mainly when she eats the cookie that makes her smaller. I would pare myself or be pared that I, too, might pass through the merest crack, a gap I know is there in the sky. I am looking just now for the cookie. Sometimes I open, pried like a fruit. Or I am porous as old bone, or translucent, a tinted condensation of the air like a watercolor wash, and I gaze around me in bewilderment, fancying I cast no shadow. Sometimes

I ride a bucking faith while one hand grips and the other flails the air, and like any daredevil I gouge with my heels for blood, for a wilder ride, for more.

There is not a guarantee in the world. Oh your *needs* are guaranteed; your needs are absolutely guaranteed by the most stringent of warranties, in the plainest, truest words: knock; seek; ask. But you must read the fine print. "Not as the world giveth, give I unto you." That's the catch. If you can catch it it will catch you up, aloft, up to any gap at all, and you'll come back, for you always come back, transformed in a way you may not have bargained for—dribbling and crazed.

The waters of separation, however lightly sprinkled, leave indelible stains. Did you think, before you were caught, that you needed, say, life? Did you think you would keep your life, or anything else you love? But no, your needs are all met. Just not as the world giveth. You see the needs of your own spirit met whenever you have asked, and you have learned that the outrageous guarantee holds. You see creatures die, and you know you will die. And one day it occurs to you that you must not need life. Obviously. And then you're gone. You have finally understood that you're dealing with a maniac.

I think that the dying pray at the last not "please," but "thank you," as a guest thanks his host at the door. Falling from airplanes the people are crying thank you, thank you, all down the air; and the cold carriages draw up for them on the rocks. Divinity is not playful. The universe was not made in jest but in solemn, incomprehensible earnest. By a power that is unfathomably secret, and holy, and fleet. There is nothing to be done about it, but ignore it, or see. And then you walk fearlessly, eating what you must, growing wherever you can, like the monk on the road who knows precisely how vulnerable he is, who takes no comfort among death-forgetting men, and who carries his vision of vastness and might around in his tunic like a live coal which neither burns nor warms him, but with which he will not part.

TSUNAMI

ON APRIL 30, 1991—on that one day—138,000 people drowned in Bangladesh. At dinner that night I brought it up. My daughter was seven years old.

I said it was hard to imagine 138,000 people drowning.

"No, it's *easy,*" my daughter said. "Lots and lots of dots, in blue water."

She was young. The tsunami victims in Bangladesh, and later, on the shores of the Indian Ocean, in Japan, they weren't dots. They were beloved daughters, beloved sons. Partners in love, fathers and mothers. Every adult knows this.

It's been a stunning time for us adults. It always is. Nothing is new, but it's fresh for every new crop of people. What is eternally fresh is our grief. What is eternally fresh is our astonishment. What is eternally fresh is our question: What the *Sam Hill* is going on here?

Is anyone running this show? Is some Fate carefully placing earthquakes on our one planet? Does an intelligence fix the height, speed, and angle of waves? Does

Omnipotence hurl hurricanes, point tornadoes, plant plagues? We could not find anyone to make a credible case for any of these brain-snarling positions.

After all, we in the West hold the individual precious. *Do we not?* Or does an individual's value weaken with the square of the distance, like the force of gravity?

We eat at restaurants while people weaken and starve everywhere, sons or daughters all. We vote as equal persons. Some monks train themselves out of bias for family members. Monks don't have children. You and I, then, are just two of seven-plus billion people of supreme significance. "Head-Spinning Numbers Cause Minds to Go Slack," read one newspaper headline. Surely we agree that our minds must not go slack.

A British journalist, observing the Sisters of Charity in Calcutta, reasoned: "Either life is always and in all circumstances sacred, or intrinsically of no account; it is inconceivable that it should be in some cases the one, and in some the other." For "sacred" substitute "of great value" or whatever you want, and look for flaws in his logic. He meant, of course, *human* life.

We who breathe air now will join the vast layers of those who breathed air once. We arise from dirt and dwindle to dirt, and the might of the universe is arrayed against us.

FOOTPRINTS

ON THE DRY LAETOLI PLAIN OF NORTHERN TANZANIA, Mary Leakey found a trail of hominid footprints. The two barefoot prehumans walked closely together. They walked on moist volcanic tuff. We have a record of those few seconds from a day about 3.75 million years ago—before hominids chipped stone tools. Ash covered the footprints and hardened like cement. Ash also preserved the pockmarks of the raindrops that fell beside the two who walked; it was a rainy day. We have almost ninety feet of their steady footprints intact. We do not know where they were going or why. We do not know why one of them paused and turned left, briefly, before continuing. "A remote ancestor," Leakey guessed, "experienced a moment of doubt." We do know we cannot make anything so lasting as these two barefoot ones did.

After archaeologists studied this long strip of record for several years, they buried it again to save it. Along one preserved portion, however, new tree roots are already cracking the footprints, and in another place winds threaten to sand them flat; the preservers did not cover them deeply enough. Now they are burying them again.

Giacometti said, "The more I work, the more I see things differently; that is, everything gains in grandeur every day, becomes more and more unknown, more and more beautiful. The closer I come, the grander it is, the more remote it is."

FOR THE TIME BEING

JUNE 1923: The French paleontologist Teilhard de Chardin was traveling on muleback in the vastness beyond the Great Wall, west of what was then called Peking. He saw from the mule what he had often seen in Egypt years before: "the burnt stones of the desert and the sand of the dunes in the dusk." This was the Ordos, the Inner Mongolian Desert.

The Ordos is a desert plateau—3,000 feet high, spreading 35,000 square miles—from which mountains rise. The Great Wall separates the Ordos from the fertile lands to the east and south in Shansi and Shensi Provinces.

He was forty-two years old, tall and narrow, fine-featured. He wore a big felt hat like a cowboy and heavy boots. Rough weather had cut lines on his face. He had carried stretchers during World War I for a regiment of sharpshooters. His courage at the front—at Ypres, Arras, and Verdun—won him several medals which the surviving men of his regiment requested for him. One of his fellows recalled his "absolute contempt for danger" as he mounted parapets under fire. They shortened his name—Pierre Teilhard de Chardin—to simply Teilhard, "Tay-YAR," as it's pronounced in French.

His characteristic expression was simple and natural, according to one scientist, who also noted that his eyes were "filled with intelligence and understanding." Another colleague described him as "a man of self-effacing and irresistible distinction, as simple in his gestures as in his manners. . . . His smile never quite turned to laughter. . . . Anxious to welcome, but like a rock of marble." From the back of a jog-trotting mule, he could spot on stony ground a tiny rock that early man had chipped.

On some days in the Ordos, he and his geologist partner dug, excavated, and sifted the ground. On other days they moved in caravan. They rode with two Mongolian soldiers—to fight bandits—and five so-called donkey boys. "On the third day," he wrote a friend, "we arrived at an immense steppe over which we traveled for more than six days without seeing much else but endless expanses of tall grasses." He passed the garnet and marble gorges of the Ula-Shan, "the old crystalline shelf of China."

July 1923: Teilhard was one of the men who unpacked the expedition's three donkeys and ten mules for the night. Bandit raids had routed them from the steppes and forced them to enter the badlands. That night he and the others pitched their two white tents in the Ordos massif, within a circle of red earth cliffs. In one red cliff he found, by daylight, the fossil remains of extinct pachyderms from the Pliocene.

"The immense hazard and the immense blindness of the world," he wrote, "are only an illusion."

The scant rain that reaches the Ordos falls in thunderstorms. During one storm, Teilhard wrote a letter. "Of this part of the journey, the crossing of the Arbous-Ula will remain in my memory as the finest stage. The innumerable strata of this savage mountain, a forward bastion of the Ula-Shan on the right bank of the Yellow River, end gently into two long concentric folds which seem to unfurl over the eastern solitudes."

August 1923: Once more they pitched their tents in the desert, in a circle of cliffs. They camped for a month, in the southeast corner of the Ordos, where the cliffs were gray, yellow, and green. Here the great eroded loess hills met the sands laid by the river called Shara-Osso-Gol. And here they found the world's first evidence of pre-Neanderthal man in China. (People lived in China long before Neanderthals lived in Europe.) The man of the yellow earths, Teilhard named him, for loess is fine yellow dust. They found his traces in the Shara-Osso-Gol's twisted canyon.

First they struck Neanderthal tools ten meters down: scrapers, gravers, quartzite blades. Then they dug through sixty-four feet of sand before they revealed an ancient hearth where Paleolithic people cooked. Their blackened hearth near the river made a thin layer among cross-bedded dune sands and blue clays. No hominid bones were there, but some tools lay about, and the hearth was indisputable—the first trace of human life north of the Himalayas.

The people who made these fires by this river about 450,000 years ago, before the last two ice ages, were not *Homo sapiens*. They were *Homo erectus,* or Peking Man. During their time, the Outer Mongolian plateau to the north continued its slow rise, blocking Indian Ocean monsoons; the northern plateau dried to dust and formed the Gobi Desert. The people would have seen dust clouds blow from the north, probably only a few big dust clouds every year. *Such dust today!* they must have thought. After the people vanished, the dust continued to blow across their land; it laid yellow and gray loess deposits hundreds of feet deep. Almost 4,500 centuries passed, and in 1222 Genghis Khan and his hordes rode ponies over the plateau, over these hundreds of feet of packed loess, over the fecund dust and barren sand, over the animal bones, the chipped blades, and the hearth. Teilhard thought of this, of Genghis Khan and the ponies. "Much later," he wrote, "Genghis Khan crossed this plain in all the pride of his victories." At that time the Mongols made stirrups and horseshoes from wild-sheep horns.

Teilhard found a twentieth-century Mongol family living in the Shara-Osso-Gol canyon. Their name was Wanschock. The father and his five sons helped Teilhard excavate during the weeks he camped. The Wanschocks rode horses, kept goats, and lived in a cave scooped out of a cliff in the loess. They taught their toddlers to ride by mounting them on sheep. "The Mongols wear long hair," Teilhard wrote, "never take off their boots, are never out of the saddle. The Mongol women look you straight in the eyes with a slightly scornful air, and ride like the men."

"Throughout my whole life," he noted later, "during every minute of it, the world has been gradually lighting up and blazing before my eyes until it has come to surround me, entirely lit up from within."

Why is there sand in deserts? Where does it come from? Why is there sand on beaches? I always thought ocean waves made sand on the seashore: Waves pounded continents' rock and shattered it to stone, gravel, and finally sand. This, I learned, is only slightly true.

Lichens make more sand than ocean waves do, as do ice salt crystals. On mountaintops and on hillsides you see cracked rock faces and boulders. Lichens grow on them, in rings or tufts, secreting acids that break down the minerals. First the lichens widen cracks in the rock, then growing salt crystals split them further, until finally freezing water shatters them.

Glaciers make some sand: Their bottoms pluck boulders and stones that scour all the land in their paths. When glaciers melt, they leave behind outwash plains, boulders, rocks, gravels, sand, and clay—which is sand ground down to floury powder. Winds lift the sand and bear it aloft.

Mostly, the continents' streams and rivers make sand. Streams, especially, and fast rivers bear bouncing rocks that knock the earth, and break themselves into sharp chips of sand. The sand grains leap—saltate—downstream. This is why the banks and the bottoms of most streams are sandy. Look in any small stream in the woods or mountains, as far inland as you like. That stream is making sand, and sand lies on its bed. Caddisfly larvae use it as stones for their odd masonry houses.

Rivers bear sand to the sea. As rivers slow, they drop

their sand, and harbors silt up and deltas spread. If the land's rock is fresh lava, as it is in Tahiti and on the Caribbean coast of Costa Rica, then the sand the streams bear down to the beaches is black. If the inland rock is basaltic, like the Columbia River plateau's, the sand the river carries to beaches is dark and fine. If the rock is granite, as it is in the eastern United States, the sand is pale quartz and feldspar.

When Los Angeles and Orange Counties dammed their intermittent streams, all the beaches from Los Angeles to Newport Beach lost their sand supply. Those weak hillside streams, which had never even flowed year-round, had supplied all that sand. Now beach towns buy up dredged harbor sand to ship and dump on their coasts to make beaches.

Coastal currents smear sand round the continents' edges, but except where waves beat cliffs, ocean waves do not make stony sand. Mostly, waves and longshore currents spread river sand coastwise, and waves fling it back at the continents' feet. Ocean waves crumble dead coral reefs. And parrot fish eat coral polyps. The fish do not digest the corals' limey bits, but instead defecate them in dribbles, making that white coral sand we prize on tropical beaches. Little or no sand lies under the deep oceans.

So why is there sand in deserts? Because windblown sand collects in every low place, and deserts are low, like beaches. However far you live from the sea, however high your altitude, you will find sand in ditches, in roadside drains, and in cracks between rocks and sidewalks.

Sand collects in flat places, too, like high-altitude deserts. During interglacials, such as the one in which we live now, soils dry. Clay particles clump and lie low, while sand grains part and blow about. Winds drop sand by weight, as one drops anything when it gets too heavy for one's strength. Winds carry light stone dust—loess—far afield, though it stays put in only a few places: in the rich prairies in central North America, and in precious flat basins of China and Russia.

September 1923: They rode back into Peking. Their mules carried 5,600 pounds of fossils and rocks in sixty wooden crates. Teilhard carried a notebook in which he had written, among other things, a morning prayer: "Be pleased yet once again to come down and breathe a soul into the newly formed, fragile film of matter with which this day the world is to be freshly clothed."

The realm of loose spirit never interested Teilhard. He did not believe in it. He never bought the view that the world was illusion and that spirit alone was real. He had written in his notebook from a folding stool in the desert of the Ordos, "There are only beings, everywhere."

Matter he loved: people, landscapes, stones. Like most scientists, he was an Aristotelian, not a Platonist. When he was still in college, he published articles on the Eocene in Egypt and the minerals of Jersey. In his twenties he discovered a new species of fish, and a new owl. His major contributions to science came after this Ordos trip, when he dated Peking Man and revised the geology of all the Quaternary strata not only through China and Mongolia but also through Java, India, and Burma. He spent twenty-three years of his adult life far from home in China, almost always in rough conditions. Why knock yourself out describing a dream?

"If I should lose all faith in God," he wrote, "I think that I should continue to believe invincibly in the world."

Teilhard had glimpsed the Gobi Desert from muleback on his 1923 Ordos expedition. It was the biggest desert on earth: 500,000 square miles of sandstorms and ravaged plateaus in what was then northern Mongolia. "As far as the eye could see around us, over the vast plain which had once been leveled by the Yellow River, waved the grass of the steppes." The solitudes moved him: the "wide torrential valleys where herds of gazelles could be seen, nose to wind, among the pebbles and the sparse grass. . . . We were crossing the low steppes of

San-Tao-Ho. The Mongolians are now no longer here. . . . The season of the yellow winds is over."

The next morning, he broke camp by the waters of the Shiling-Gol and moved toward Kalgan in the Gobi, an area science did not know. He found fossils. Two days later, he was wielding a pick at the Dalai-Nor, a wet salt pan twenty-five miles long on the Mongolian steppe. He shook and spread his bedroll on a dune by the shore. Six oxcarts carried supplies and boxes of extinct Tertiary horse and rhino bones.

In the field Teilhard wore a tough jacket and a wide-brimmed slouch hat. In one breast pocket he carried a breviary, and in the other a pack of Gauloises. "This man with the clear regard," a friend called him. He was long-boned, sharp-faced, faintly smiling when serious, and merry in company. When he laughed his face split into planes. His friends were mostly geologists, paleontologists, priests, explorers, educated Paris and New York women, and archaeologists. Among them were an odd trio: Julian Huxley, Henry Clay Frick, and Paul Valéry.

The paleontologist once called God "punctiform": "It is precisely because he is so infinitely profound and punctiform that God is infinitely near." Is it useful and wise to think of God as punctiform? I think so.

Of the gospel miracles Teilhard wrote, "I feel obliged

to admit that I believe not because of but in spite of the miracles."

The more nearly spherical is a grain of sand, the older it is. "The average river requires a million years to move a grain of sand one hundred miles," James Trefil tells us. As a sand grain tumbles along the riverbed—as it saltates, then lies still, then saltates for those millions of years—it smooths some of its rough edges. Sooner or later, it blows into a desert. In the desert, no water buoys its weight. When it leaps, it lands hard. In the desert, it knaps itself round. Most of the round sand grains in the world, wherever you find them, have spent some part of their histories blowing around the desert.

"We live surrounded by ideas and objects infinitely more ancient than we imagine," Teilhard said, "and yet at the same time everything is in motion."

Chert, flint, agate, and glassy rock can flake to a cutting edge just a few atoms thick. Prehistoric people made long oval knives of this surpassing sharpness, and made them, wittingly, too fragile to use. Those people—*Homo sapiens*—lived in a subfreezing open-air camp in central France about 18,000 years ago. We call their ambitious culture Solutrean; it lasted only about 3,000 years. It was they who invented the bow and arrow, the spear thrower, and the needle—which made clothes such a welcome improvement over draped pelts.

Solutrean artisans knapped from rocks astonishing yellow blades in the shape of long, narrow, pointed leaves. Most of these blades are the size and thickness of a fillet of sole. The longest is fourteen inches long, four inches at its beam, and only one-quarter inch thick. Their intricate technique is, according to Douglas Preston, "primarily an intellectual process." A modern surgeon at Michigan Medical School used such a blade to open a patient's abdomen; it was smoother, he said, than his best steel scalpels. Another scientist estimated a Solutrean chert blade to be one hundred times sharper than a steel scalpel; its edge split few cells, and left scant scar. An Arizona rancher skinned a bear with an obsidian knife in two hours, he said, instead of the usual three and a half; he never even needed to press down.

Hold one of these chert knives to the sky. Most of it shines dull, waxy gold—brown in the center, and yellow toward the edges. At each fractured rim, however, the blade thins from translucency to transparency. You see your skin, and the sky. At its very edge the blade dissolves into the universe at large. It ends, imperceptibly, at an atom.

Each of these delicate, absurd objects takes hundreds of separate blows to make. At each stroke and at each pressure flake, the brittle chert might—and, by the record, very often did—snap. The maker knew he was likely to lose many hours' of breath-holding work at a single tap. The maker worked in extreme cold. He knew no one would ever use the virtuoso blades. He protected them, and his descendants saved them intact, for their very perfection. To any human on earth, the sight of one of them means: Someone thought of making, and made, this difficult, impossible, beautiful thing.

New sand is young and sharp. Some of the sand in a sidewalk crack will cut your finger. The geologist Philip H. Kuenen, who devoted his working life to sand, reckoned, perhaps imprecisely, that every second, one billion sharp new sand grains of quartz alone appear on earth, chips off the old continental blocks. Sand has been forming at this clip all along. Only a smattering of that sand ends up on beaches and deserts. So why are we all not buried in dunes? Because sand amasses in basins whose floors subside. Pressure cooks much of it into sandstone, as one crustal plate slides over the next like a hand.

Exposed uplifted sandstone, naturally, can wear away again. A sandstone castle in Austria, 900 years old, is itself returning to soil. Weathering has turned its outer walls to fine clay from which grass grows.

Sand grains bang about in deserts and wear down their angles. Kuenen went so far as to determine how much desert the world "needs"—2×10^6 square kilometers—in order, as *Sand and Sandstone* explained it, "to keep the world average roundness constant (to offset the new, sharp-cornered sand added each year)." So you can easily reason that if erosion and drought fail to form new deserts in Africa, say, at an acceptable pace, thereby starving whole populations, the ratio of the world's round sand to the world's sharp sand will get seriously out of whack.

Teilhard de Chardin was a Jesuit priest as well as a paleontologist. The theology and cosmology that drove his thinking and writing are not his strongest legacy, any more than William Butler Yeats's theology and cosmology are his. He wrote eighteen books. The unhappy prominence of his dull, arcane, and improbably crackpot *The Phenomenon of Man* thirty years ago, and the occasional nutty enthusiasm of his admirers, some of them vague-brained new-agers, have obscured his intelligent, plausible, and beautiful *The Divine Milieu* and the short, magnificent literary essays "The Mass of the

World" and "The Heart of Matter." The world rarely can or will distinguish art from mere opinion. Pressed for his opinions, Teilhard produced them, and their peculiarly disagreeable lexicon. The cranks they attracted possibly tempted some possessors of good minds to write him off without reading him.

In France he had taken years of theology courses, and admitted that he did not find them *bien amusants*. He studied chemistry and physics in Cairo; at the Sorbonne he worked in botany and zoology as well as geology. His doctorate in geology described mammifers of the Lower Eocene in France.

He ran afoul of Roman authorities over his thinking. In the 1920s, evolution was still a new current in thought, as the church reckoned, and it had not yet penetrated Rome's layers of brocade. To the church, the notion of biological evolution seemed to hash the old doctrine of original sin. After Teilhard lectured on evolution in Paris, the church in Rome gagged him. It forbade him to lecture and to publish anything but purely scientific articles. He complied. Of his eighteen books, the church permitted only one to see light in his lifetime, a short scientific monograph published in Peking. The cardinals were pleased to keep his person, also, tucked away. They exiled him to China, the second time for virtually the rest of his life. He was forty-two. Always longing for France, for his Paris teaching position, the Jesuit brothers, and his friends, and always willing to settle for a life in the United States, he nevertheless discovered gradually that his vow of obedience would require him to renounce the West for twenty-two years more.

Every year, he applied to publish his work; every year, Rome refused. Every year, he applied to return to France; every year, Rome refused. At last Rome let him visit France briefly when he was sixty-five; he had had a heart attack. Still Rome prohibited his publishing. Offered a fine teaching post, he went to Rome in person to seek permission; Rome denied it. He traveled to the United States, to South America, and to Africa, and he visited Paris to spread his ideas by talking. Even when he was seventy-three and dying of heart disease in New York, Rome forbade his publishing, lecturing, or returning to France.

Why did he put up with it? One of his colleagues said he had "the impatience of a prophet." When exactly did he show this impatience? His colleagues and many of his friends urged him to quit the Jesuits. Only for a few weeks, however, did he ever consider leaving the order. To kick over the traces, he thought, would betray his Christianity. People would think—perish the thought—he was straying from the church! His brother Jesuits defended him and his thinking. Leaving the order would mean, he decided, "the killing of everything I want to liberate, not destroy." The Catholic Church, he wrote late in life, is still our best hope for an arch to God, for the transformation of man, and for making, in his view, evolution meaningful: It is "the only international organization that works."

Again and again he had dedicated his life wholeheartedly to the church: Consequently, he did not much complain. When he first learned that Rome had blocked publication of *The Divine Milieu,* he did, however, allow himself to write a friend in private that it was "a pity." The year before he died, while he was declaring in sincere letters that Rome was mankind's best hope, he also allowed himself to blow off steam, like many a cleric. "The sin of Rome," he wrote to a friend, "is not to believe in a future. . . . I know it because I have stifled for fifty years in this sub-human atmosphere." He apparently felt strongly both ways. Later, Vatican II calmly endorsed most of his ideas.

"All that is really worthwhile is action," Teilhard wrote. "Personal success or personal satisfaction are not worth another thought."

Earth sifts over things as dirt or dust. If you stay still, earth buries you, ready or not. The debris on the tops of your feet or shoes thickens, windblown dirt piles around it, and pretty soon your feet are underground. Then the ground rises over your ankles and up your shins. If the sergeant holds his platoon at attention long enough, he and his ranks will stand upright and buried like a Chinese emperor's army.

Micrometeorite dust can bury you, too, if you wait: A ton falls on earth every hour. Or you could pile up with locusts. At Mount Cook in Montana, at 11,000 feet, you can see on the flank a dark layer of locusts. The locusts fell or wrecked in 1907, when a swarm flew off course and froze. People noticed the deposit only when a chunk separated from the mountain and fell into a creek that bore it downstream.

The rate at which dirt buries us varies. New York City's street level rises every century. The Mexico City in which Cortés walked is now thirty feet underground. It would be farther underground except that Mexico City itself has started sinking. Digging a subway line there, workers found a temple. Debris lifts land an average of 4.7 feet per century. King Herod the Great rebuilt the Second Temple in Jerusalem 2,000 years ago. The famous Western Wall is a top layer of old retaining wall near the peak of Mount Moriah. From the present bottom of the Western Wall to bedrock is sixty feet.

Quick: Why aren't you dusting? On every continent,

we sweep floors and wipe tabletops not only to shine the place but to forestall burial.

It is interesting, the debris in the air. A surprising portion of it is spider legs, and bits thereof. Spider legs are flimsy, Oxford writer David Bodanis says, because they are hollow. They lack muscles; compressed air moves them. Consequently, they snap off easily and blow about. Another unexpected source of aerial detritus is tires. Eroding tires shed latex shreds at a brisk clip, say the folks who train their microscopes on air. Farm dust joins sulfuric acid droplets (from burned fossil fuels) and sand from the Sahara Desert to produce pollution haze that blurs and dims the world.

We inhale "many hundreds of particles in each breath we take," says Bodanis. Air routinely carries intimate fragments of run, dung, carcasses, leaves and leaf hairs, coral, coal, skin, sweat, soap, silt, pollen, algae, bacteria, spores, soot, ammonia, and spit, as well as "salt crystals from ocean whitecaps, dust scraped off distant mountains, micro bits of cooled magma blown from volcanoes and carried microfragments from tropical forest fires." These things can add up.

At dusk the particles meet rising water vapor, stick together, and fall: That is when they will bury you. Soil bacteria eat what they can, and the rest of it stays put if there's no wind. After thirty years, there is a new inch of topsoil, which may, however, wash into the ocean.

We live on dead people's heads. Scratching under a suburb of St. Louis, archaeologists recently found thirteen settlements, one on top of the other, some of which lasted longer than St. Louis has. Excavating the Combe Grenal cave in France, paleontologists found sixty different layers of human occupation.

The pleasantly lazy people of Bronze Age Troy cooperated with the burial process. Instead of sweeping garbage and litter from their floors, they brought in dirt to cover the mess and tramped it down. Soon they had to stoop in their rooms, so they heightened their doors and roofs for another round. Invaders, too, if they win, tend to build new floors on roofs they've wrecked. By the nineteenth century, archaeologists had to dig through twenty-four feet of earth to find the monuments of the Roman Forum.

In 1870–71 when Heinrich Schliemann was digging at a site he hoped was Troy, he excavated a trench sixteen feet deep before he found worked stones. He had found the top of a wall twenty feet high. Under that wall's foundation, he learned over years of digging, was another high wall, and—oops—another, and another. Archaeologists are still excavating Troy.

Elsewhere, the ziggurats of the ancient New East sank into the ground, settled into soft soils, and decomposed. "Every few years, the priests would have them build up a few steps higher to compensate for the sinking of the bottom story into the soil." Earthworm tunnels lower buildings, too, as Darwin noticed. These days the heavy Cathedral of St. John the Divine in New York City is sinking, according to the cathedral's recent writer-in-residence William Bryant Logan, who wrote the excellent book *Dirt*. The cathedral's base "is now beneath the water table," and "a living spring" has risen in its crypt.

In Santa Monica, California, early every morning, a worker in a bulldozer plows the previous day's trash into the beach. I saw it. He turns the trash layer under as a farmer lashes fields with last year's leaves. He finishes the top by spreading a layer of sand, so the beach, rising on paper and Styrofoam, looks clean.

When he entered the war, Teilhard was already a priest. One dawn in 1918, camped in a forest in the Oise with his Zouave regiment, he had neither bread nor wine to offer at Mass. He had an idea, however, and he wrote it down.

Five years later, he sat on a camp stool inside a tent by the Ordos desert cliffs west of Peking. He reworked his old wartime idea on paper. What God's priests, if empty-handed, might consecrate at sunrise each day is that one day's development: all that the evolving world will gain and produce, and all it will lose in exhaustion and suffering. These the priest could raise and offer.

In China again, four years later yet, he rode a pony north into the Mongolian grasslands and traced Quaternary strata. Every day still he said to himself what he now called his Mass upon the altar of the world, "to divinize the new day." "Since once more, my Lord, not now in the forests of the Aisne but in the steppes of Asia, I have neither bread, nor wine, nor altar, I shall rise beyond symbols to the pure majesty of the real, and I shall offer you, I your priest, on the altar of the whole earth, the toil and sorrow of the world."

Sand plunges. Sandstone plates subduct. They tilt as if stricken and die under crusts. At abyssal depths the earth's weight presses out their water; heat and weight burst their molecules, and sandstone changes into quartzite. It keeps the form of quartzite—that milky gray mineral—to very great depths, where at last the quartzite melts and mixes into magma. In the fullness of time, magma rises along faults; it surfaces, and makes the continents that streams will one day grate back to sand.

"I feel no special assurance of the existence of Christ," Father Teilhard explained cheerfully at the end of a book in which he tracked his ideas. His evolving universe culminates in Christ symbolically. "As much as anyone, I imagine," he went on, "I walk in the shadows of faith"—that is, in doubt. Doubt and dedication, after all, go often hand in hand. And "faith" crucially, is not assenting intellectually to a series of doctrinal propositions; it is living in a conscious and rededicated relationship to God. Nevertheless, the temptation to profess creeds with uncrossed fingers is strong. Teilhard possessed, like many spiritual thinkers, a sort of anaerobic capacity to thrive on paradox.

It was in 1928, when Teilhard was forty-seven, that his team discovered a bone from Peking Man. His partner, a Chinese archaeologist, found a man's skull. Years before, Teilhard had unearthed the first tools, and the first hearth; now here were the first bones—the first to be found in all of Asia. Time had stuffed the skull down a red fissure in a blue cave wall at Zhoukoudian, near Peking. It was then that the team named this "Peking Man."

They found the cave by questioning a big-city pharmacist. Many old folk in China, even today, drink suspended-fossil-bone powders as elixirs—so-called dragon's teeth elixirs. Consequently, paleontologists for many generations have checked Chinese pharmacies and asked, "Where exactly did these bones come from?" In this way, one such specialist, shopping for fossils, recognized a human tooth, and his inquiry led to the caves at Zhoukoudian—Dragon Bone Hill.

Hauling his camp cot from Peking, Teilhard lived with villagers as he directed the dig. Over the years he sorted and eventually named the fissures' many other animal bones. He discovered bones from saber-toothed tigers, ostriches, horses, a large camel, buffalo, wild sheep, rhinos, hyenas, and "a large and small bear."

Ultimately he was able to date Peking Man in the Pleistocene. He established the date by various methods, one of which was interesting: Among the bits of debris under, around, and above various layers of the hominid's bones and tolls were skulls, whole or in fragments, of mole rats. He undertook his own study of the mole rats' evolving skulls, dated them, and so helped confirm Peking Man's dates.

The team dug further into the immensities of the Zhoukoudian caves; for ten years they excavated, for eight months a year. Teilhard retrieved five more human skulls, twelve lower jaws, and scattered teeth. It was his major life's work.

During those ten years, squinting and laughing furrowed his face. His temples dipped as his narrow skull bones emerged. When he could not get Gauloises, he smoked Jobs. Daily he said the Divine Office—the liturgy, mostly psalms, that is the prayer of the Catholic (and Anglican, Episcopal, Presbyterian, and Lutheran, for that matter) Church. A British historian who knew him described his "kindly and ironic grace," his "sharp and yet benevolent refinement."

In all those years, he found no skeletons. When colleagues worldwide praised him for his discoveries, Teilhard spoke with modesty, and even exasperation: "Heads," he said, "practically nothing but heads."

A hundred years later, after several decades' chaos halted the work, paleontologists from all over the world are again finding hominid bones in the Zhoukoudian caves, along with choppers and stone flakes.

Peking Man and his people walked upright; with limbs like ours they made fire and stone tools. The land was jungly then. They ate mostly venison and hackberries. They hunted elephants, tigers, and boars. They lived before water filled the Great Lakes, before the Florida peninsula rose from the sea, while camels and mastodons grazed across North America. They lived before the two great ages when ice covered Scandinavia and Canada, as well as the British Isles, northern Germany, and the northern United States. They lived before the Atlantic Ocean drowned eastern North America between glaciations. Their human species is now extinct.

Most paleontologists believe that we—we humans in the form of *Homo erectus*—left Africa 90,000 years ago by walking up the Great Rift Valley, generation after generation, to the valley's end at the Sea of Galilee. Recent, much older *erectus* finds in Java, China, and the Republic of Georgia seem to show, however, that our generations started leaving Africa about a million years earlier—unless, that is, modern humans arose in Asia. These newer, more ancient dates jolt paleontologists, who one might expect to be accustomed to this sort of thing by now—this repeated knocking out of the back wall, this eerie old light cast on the long-peopled landscape.

Whenever we made our move, we did not rush to Corfu like sensible people. Instead we carried our cupped fires into the lands we now call the Levant, and then, seriatim, into China, Japan, and Indonesia, whence we hopped islands clear to Australia. There, on a rock shelter, we engraved animals twice as long ago as we painted cave walls in France. In other words, people—*erectus* included—plied the Asian islands thousands of years before Europe saw any humans who could think of such a thing as a raft.

"However far back we look into the past," Teilhard said, "we see the waves of the multiple breaking into foam."

During the violence and famine caused by the Japanese invasion of China, that first Peking Man skull disappeared from the Chinese museum in which it was housed. Scientists suspect starving locals pulverized and drank it. Remaining, however, was a plaster cast of this skull, as there were casts of every bit of bone and tooth—forty people's remains—that the team found by working the site for all those years. Those plaster casts have proved handy, since every single one of the actual Peking Man bones, crate after crate, disappeared in World War II. Scientists cached the crates with a U.S. Marine physician who tried to carry them back as luggage. The Japanese caught him, though before he went to prison camp, he was able to entrust the crates to European officials and Chinese friends. Unfortunately, when he left prison four years later, after the war had ended, the crates had disappeared. Recent searches draw only blanks.

The man of the red earths, Teilhard called Peking Man. And of Christianity he said, "We have had too much talk of sheep. I want to see the lions come out."

Early spring 1930: Father Teilhard, wearing his clerical collar, was having afternoon tea in the Peking courtyard garden of his new friend, an American woman named Lucile Swan. He sat erect and relaxed on a bamboo chair at a rattan table, laughing and talking. We have a snapshot. In the other bamboo chair Lucile Swan turned his way; she looked mightily amused. A headband held her short, curly hair from her firm and wide-boned face. She wore an open parka and pants: It was perhaps chilly for taking tea outdoors. Her small dog, white and brown, sat at her knee watching the merriment, all ears.

He was forty-nine then; she was forty, a sculptor, divorced. It was more than a year since the Peking Man discovery. Teilhard was living in a village near the Zhoukoudian cave and coming into Peking once a week. The two had met at a dinner party and liked each other at once. "For the first time in years I felt young and full of hope again," she recalled. She had attended an Episcopal boarding school in Iowa and, later, the Art Institute of Chicago. In Peking, she made portrait sculptures in clay and bronze, and groups of semi-abstract figures; throughout her life she exhibited widely.

Soon the two established a daily routine in Peking: They walked, took tea at five, and he returned across the city to the Jesuit house at six. Those first several years, they laughed a great deal—about, among many other things, the American comic *The Little King,* which Lucile found in her *New Yorkers* and translated for him. Their laughter carried over courtyard walls.

"Lucile was fine-featured, amply bosomed," a friend who joined them at tea would later remember, "beloved by all who knew her, for she glowed with warmth and honest sentiment." And Father Teilhard was "a lean, patrician priest . . . the jagged aristocrat. He radiated outward, gravely, merrily, inquiringly. And always with a delicate consideration for the other and no concern for self."

June 1930: "Our blue tents are pitched at the edge of a fossil-bearing cliff looking out over the immense flat surface of Mongolia," he wrote. "We work in solitude." He knew he could not post this letter for several months, for he was tracing the wild bounds of Outer Mongolia. "Cut off from any correspondence, I feel that my Paris hopes are dormant." He was not yet writing letters to Lucile Swan.

He had interrupted his Zhoukoudian caves dig to join an American expedition: the 1930 Roy Chapman Andrews expedition. Most of his past five years he had already spent traveling with mules to dig the great Gobi marches; this new adventure would take him even farther afield. To fix Peking Man in context, he hoped to discover the geological history of the Quaternary through all of Asia. In the course of the expedition's wild and crawling journey, which lasted most of a year, he would, in fact, find the evidence necessary to link and date Chinese and Mongolian strata.

The Andrews expedition was a step up for the *monsieur* accustomed to mules. They drove Dodge trucks. Strings of camels carried gas. Digging, they encountered between five and ten poisonous brown pit vipers every day. The vipers kept them alert, one team member reported. Characteristically, Teilhard never mentioned them in his letters. He liked Roy Chapman Andrews, who made his name finding dinosaur eggs. "A wonderful talker," Teilhard described the expedition's leader, and a hunter who, when the team lacked food, drove off into the bright expanses and returned "with a couple of gazelles on the running boards." Teilhard's own vitality still battened on apparent paradox. The man who said that his thirty months on the front in the Great War had made him "very mystical and very realistic," now wrote from his blue tent in Mongolia that "rain, storms and dust and icy winds have only whipped up my blood and brought me rest." They called the place Wolf Camp, for the wolves that, along with eagles, hunted there.

"Purity does not live in a separation from the universe," he wrote, "but in a deeper penetration of it."

The next year he attached himself to a rough French expedition as its geologist. The 1931 Croisière Jaune expedition took nine months and crossed Asia to the Russian frontier. Teilhard doubled his knowledge of Asia. He went so far west that he realized one day he was halfway from Peking to Paris. He and the other Frenchmen traveled by Citroën caterpillar across "great folds of impassable land." They breached what he admired as "the unending corrugations of the Gobi peneplain and the monumental formations of Upper Asia." They crossed a region where mountains rose 21,000 feet. The Silk Road's northern route took them west to the Pamir Mountains as far as Afghanistan. On the road, the others reported, the paleontologist often stopped his Citroën half-track, darted ahead into the water, and picked up a chipped green rock, a paleolith, or a knob of bone.

"This vast ocean-like expanse," he wrote, "furrowed by sharp ridges of rock, inhabited by gazelles, dotted with white and red lamaseries . . . I am obliged to understand it." He examined the juncture where the foot of "the huge ridge of the Celestial Mountains" plunged 600 feet below sea level into the Turfan Deep. The Turfan Deep, in turn, opened into a "vast depression" in which the River Tarim lost itself in the shifting basin of the Lop-Nor.

"I still, you see, don't know where life is taking me," he wrote his friend Max Bégouën. "I'm beginning to think that I shall always be like this and that death will find me still a wanderer."

Returning midwinter, the Croisière Jaune team explored an immense section of the Gobi no one had mapped. The temperature stuck between –4 and –22 degrees F. They dared not let the caterpillars' engines stop. Twice a day they halted and stood by the mess vehicle, nearly immobile in furs, and tried to down the boiling soup in their tin mugs before it froze.

By the time he was fifty, Teilhard said, he had awakened to the size of the earth and its lands. In only his first ten years there, he had explored China at walking pace, from the Pacific to Afghanistan, and from the Khingan Mountains northeast of Mongolia all the way south to Vietnam. Returning from the Croisière Jaune expedition, he had worked all spring in Peking, and traveled throughout the fall. It was then, in 1932, three years after meeting her, that he began writing letters to the sculptor with whom he had taken tea behind that red courtyard gate.

In his salutations, "Lucile, dear friend" quickly became "Lucile dear" and then "Dearest." She remained "Dearest" (sometimes underlined) for twenty-three years, until he died. Their published correspondence—hundreds of letters apiece—knocks one out, for of course she loved him, and he loved her. "I am so full of you, Lucile. —How to thank you for what you are for me! . . . I think that I have crossed a critical point in my internal evolution, those past months, —with you. . . . My dream," he wrote her, "is to make you gloriously happy."

She translated his work. She molded in clay for science a fleshed-out head of Peking Man. For her he sounded out his ideas. One idea he returned to quite often was his commitment to his vows. "I do not belong to myself," he told her. In an essay he wrote, "Through woman and woman alone, man can escape from isolation," but in right passion, love will be, predictably, spiritual. "Joy and union," he wrote her, "are in a continuous common discovery. Is that not true, dearest?"

He never broke any of his vows. (Both men and

women who live under religious vows agree that while communal living irritates them most, obedience is by far the toughest vow, and not, as secular people imagine, chastity. Teilhard never had to endure twenty-four-hour communal living, as monks do; still, obedience chafed him sorely, and he confided later that to maintain chastity he had, quite naturally, "been through some difficult passages.")

"It seems sometimes that I have to accept so *many* things," Lucile Swan wrote him, while in her private journal she wrote, "Friendship is no doubt the highest form of love and also very difficult." As the years passed, he lived in Peking and visited France for months on end; he traveled to South America, Burma, India, South Africa, Rhodesia, and Java. They both lived in Peking, for the most part, during the twenty-two years following their meeting, until 1941, when she moved to the United States. Missing him sometimes by a few days, she traveled in those years, and the fourteen that followed, to France, Rome, Ethiopia, Switzerland, Siam, London, and India. In 1952, when Teilhard was seventy-one years old, he moved to New York City, where Lucile was living and exhibiting. They met frequently. "We still disturb each other," he wrote her from across town. Especially disturbing to her was his new and deep friendship with another woman—also an American, a novelist.

Even three years later, after he had survived a heart attack, and after hundreds of their love letters had flown all over the world for decades, after hundreds of reunions and partings, and after hundreds of visits in New York, he wrote her that he hoped that "things" would "gradually settle emotionally." There was not much "gradually" left, though, as he died eleven days later. A snapshot of Lucile Swan outdoors in her sixties shows a magnificent beauty. A dog holds one end of a towel in its teeth, while she holds the other in her hand; the dog, looking at her face, is clearly waiting for her to do her part right. She lived another ten years after Teilhard died.

"What is born between us is forever: I know it," he wrote her. One fervently hopes so. One also hopes—at least this one does—that in heaven souls suffer fewer scruples, or better yet, none at all.

The material world for Teilhard dissolves at the edges and grows translucent. The earth is a Solutrean blade. It thins to an atom. As a young scientist, he held the usual view that the world is all material; from it spirit could not derive. Soon he inverted the terms: The world is all spirit, from which matter cannot derive, save through Christ. "Christ spreads through the universe, dissolved at the edges." This is just the sort of idiosyncratic, brilliant lexicon that drives his doctrine-minded readers mad: Christ is chert; chert is Christ. The world is incandescent. Things are "innumerable prolongations of divine being." Or, "Things retain their individuality but seem to be lighted from within and made of active, translucent flesh."

Even the purest metaphysical Taoist thinkers, the Lungman Taoists, say that people "can assist in improving the divine handiwork"—or, as a modern Taoist puts it, people may "follow the Will of the Creator in guiding the world in its evolution toward the ultimate Reality." Even Meister Eckhart said, "God needs man." God needs man, said Teilhard as well, to disclose him, complete him, and fulfill him. Teilhard's friend Abbé Paul Grenet paraphrased his thinking about God: "His name is holy, but it is up to us to sanctify it; his reign is universal, but it is up to us to make him reign; his will be done, but it is up to us to accomplish it."

"Little by little," the paleontologist himself said, "the work is being done."

AN EXPEDITION TO THE POLE

I

THERE IS A SINGING GROUP in this Catholic church today, a singing group that calls itself "Wildflowers." The lead is a tall, square-jawed teenaged boy, buoyant and glad to be here. He carries a guitar; he plucks out a little bluesy riff and hits some chords. With him are the rest of the Wildflowers. There is an old woman, wonderfully determined; she has long orange hair and is dressed country-and-western style. A long embroidered strap around her neck slings a big western guitar low over her pelvis. Beside her stand a frail, withdrawn fourteen-year-old boy and a large Chinese man in his twenties who seems to want to enjoy himself but is not quite sure how. He looks around wildly as he sings, and shuffles his feet. There is also a very tall teenaged girl, presumably the lead singer's girlfriend; a wispy soprano, she is delicate of feature, half serene and half petrified. They straggle out in front of the altar to teach us a brand-new hymn.

It all seems a pity at first, for I have overcome a fiercely anti-Catholic upbringing precisely in order to escape Protestant guitars. Why am I here then? Who gave these nice Catholics guitars? Why are they not mumbling in Latin and performing superstitious rituals? What is the Pope thinking of?

Nobody said things were going to be easy. A taste for the sublime is, after all, a greed like any other; why begrudge the churches their secularism now, when from the general table is rising a general song? Besides, in a way I do not pretend to understand, these people—all the people, in all the ludicrous churches—have access to the land.

THE LAND

The Pole of Relative Inaccessibility is "that imaginary point on the Arctic Ocean farthest from land in any direction." It is a navigator's paper point contrived to console Arctic explorers who, after Peary and Henson reached the North Pole in 1909, had nowhere special to go. There is a Pole of Relative Inaccessibility on the Antarctic continent, also; it is that point of land farthest from salt water in any direction.

The Absolute is the Pole of Relative Inaccessibility located in metaphysics. After all, one of the few things we know about the Absolute is that it is relatively inaccessible. It is that point of spirit farthest from every accessible point in all directions. Like the others, it is a Pole of the Most Trouble. It is also—I take this as given—the pole of great price.

THE PEOPLE

It is the second Sunday in Advent. For a year I have been attending Mass at this Catholic church. Every Sunday for a year I have run away from home and joined the circus as a dancing bear. We dancing bears have dressed ourselves in buttoned clothes; we mince around the rings on two feet. Today we were restless; we kept dropping onto our forepaws.

No one, least of all the organist, could find the opening hymn. Then no one knew it. Then no one could sing anyway.

There was no sermon, only announcements.

The priest proudly introduced the rascally acolyte who was going to light the two Advent candles. As we all could plainly see, the rascally acolyte had already lighted them.

During the long intercessory prayer, the priest always reads "intentions" from the parishioners. These are slips of paper, dropped into a box before the service begins, on which people have written their private concerns, requesting our public prayers. The priest reads them, one by one, and we respond on cue. "For a baby safely delivered on November twentieth," the priest intoned, "we pray to the Lord." We all responded, "Lord, hear our prayer." Suddenly the priest broke in and confided to our bowed heads, "That's the baby we've been praying for the past two months! The woman just kept getting more and more pregnant!" How often, how shockingly often, have I exhausted myself in church from the effort to keep from laughing out loud? I often laugh all the way home. Then the priest read the next intention: "For my son, that he may forgive his father. We pray to the Lord." "Lord, hear our prayer," we responded, chastened.

A high school stage play is more polished than this service we have been rehearsing since the year one. In 2,000 years, we have not worked out the kinks. We positively glorify them. Week after week we witness the same miracle: that God is so mighty he can stifle his own laughter. Week after week we witness the same miracle: that God, for reasons unfathomable, refrains from blowing our dancing bear act to smithereens. Week after week Christ washes the disciples' dirty feet, handles their very toes, and repeats, it is all right—believe it or not—to be people.

Who can believe it?

During communion, the priest handed me a wafer

that proved to be stuck to five other wafers. Resisting the impulse to help, I waited while he tore the clump into rags of wafer. Directly to my left, and all through communion, a woman on the piano was banging out the theme from *The Sound of Music.*

THE LAND

Nineteenth-century explorers set the pattern for polar expeditions: Elaborately provisioned ships set out for high latitudes. Soon they encounter the pack ice and equinoctial storms. Ice coats the deck, spars, and rigging; the masts and hull shudder; the sea freezes around the rudder, and then fastens on the ship. Early sailors try ramming, sawing, or blasting the ice ahead of the ship, before they give up and settle in for the winter. In the nineteenth century, this being "beset" in the pack often killed polar crews; later explorers expected it and learned, finally, to put it to use. Sometimes officers and men move directly onto the pack ice for safety; they drive tent stakes into the ice and pile wooden boxes about for tables and chairs.

Sooner or later, the survivors of that winter or the next, or some of them, set off over the pack ice on foot. Depending on circumstances, they are looking either for a Pole or, more likely, for help. They carry supplies, including boats, on sledges fastened to shoulder harnesses. South Polar expeditions usually begin from a base camp established onshore. In either case, the terrain is so rough, and the men so weakened by scurvy, that the group makes only a few miles a day. Sometimes they find an island on which to live or starve the next winter; sometimes they turn back to safety, stumble onto some outpost of civilization, or are rescued by another expedition; very often, when warm weather comes and the pack ice splits into floes, they drift and tent on a floe, or hop from floe to floe, until the final floe lands, splits, or melts.

In 1847, according to Arctic historian L. P. Kirwan, the American ship *Polaris* "was struck by an enormous floe. And just as stores, records, clothing, equipment, were being flung from the reeling ship, she was swept away through the Arctic twilight, with most, but not all of her crew on board. Those left behind drifted for thirteen hundred miles on an ice-floe until they were rescued, starving and dazed, off the coast of Labrador."

Polar explorers were chosen, as astronauts are today, from the clamoring, competitive ranks of the sturdy, skilled, and sane. Many of the British leaders, in particular, were men of astonishing personal dignity. Reading their accounts of life in extremis, one is struck by their unending formality toward each other. When Scott's Captain Oates sacrificed himself on the Antarctic peninsula because his ruined feet were slowing the march, he stepped outside the tent one night to freeze himself in a blizzard, saying to the others, "I am just going outside and may be some time."

Even in the privacy of their journals and diaries, polar explorers maintain a fine reserve. In his journal, Ernest Shackleton described his feeling upon seeing, for the first time in human history, the Antarctic continent beyond the mountains ringing the Ross Ice Shelf: "We watched the new mountains rise from the great unknown that lay ahead of us," he wrote, "with feelings of keen curiosity, not unmingled with awe." One wonders, after reading a great many such firsthand accounts, if polar explorers were not somehow chosen for the empty and solemn splendor of their prose styles—or even if some eminent Victorians, examining their own prose styles, realized, perhaps dismayed, that from the look of it, they would have to go in for polar exploration. Salomon Andrée, the doomed Swedish balloonist, was dying of starvation on an Arctic island when he confided in his diary, with almost his dying breath, "Our provisions must soon and richly be supplemented, if we are to have any prospect of being able to hold on for a time."

THE PEOPLE

The new Episcopalian and Catholic liturgies include a segment called "passing the peace." Many things can go wrong here. I know of one congregation in New York that fired its priest because he insisted on their passing the peace—which involves nothing more than shaking hands with your neighbors in the pew. The men and women of this small congregation, though, had limits to their endurance; and passing the peace fell beyond those limits. They could not endure shaking hands with people against whom they bore lifelong grudges. So they fired the priest and found a new one more sympathetic to their needs.

The rubric for passing the peace requires that one shake hands with whoever is handy and say, "Peace be with you." The other responds, "Peace be with you." Every rare once in a while, someone responds simply "Peace." Today I was sitting beside two teenaged lugs with small mustaches. When it came time to pass the peace I shook hands with one of the lugs and said, "Peace be with you," to which he replied, "Yeah."

THE TECHNOLOGY

The turning point in Arctic exploration was the Franklin Expedition. The expedition itself accomplished nothing, and all its members died. But its failure to return, and the mystery of its whereabouts, attracted so much publicity in Europe and the United States that thirty ships set out looking for traces of the ships and men; these search parties explored and mapped the Arctic for the first time, found the Northwest Passage which Franklin had sought, and developed a technology adapted to Arctic conditions, a technology capable of bringing explorers back alive. The technology of the Franklin expedition, by contrast, was adapted only to conditions in the Royal Navy officers' clubs in England. The Franklin expedition stood on its dignity.

In 1845, Sir John Franklin and 138 officers and men embarked from England to find the Northwest Passage across the high Canadian Arctic to the Pacific Ocean. They sailed in two three-masted barques. Each sailing vessel carried an auxiliary steam engine and a twelve-day supply of coal for the entire projected two or three years' voyage. Instead of additional coal, according to L. P. Kirwan, each ship made room for a 1,200-volume library, "a hand-organ, playing fifty tunes," china place settings for officers and men, cut-glass wine goblets, and sterling silver flatware. The officers' sterling silver knives, forks, and spoons were particularly interesting. The silver was of ornate Victorian design, very heavy at the handles and richly patterned. Engraved on the handles were the individual officers' initials and family crests. The expedition carried no special clothing for the Arctic, only the uniforms of Her Majesty's Navy.

The ships set out in great glory and fanfare. Franklin uttered his utterance: "The highest object of my desire is faithfully to perform my duty." Two months later a British whaling captain met the two barques in Lancaster Sound; he reported back to England on the high spirits of officers and men. He was the last European to see any of them alive.

Years later, civilization learned that many groups of Inuit had hazarded across tableaux involving various still-living or dead members of the Franklin expedition. Some had glimpsed, for instance, men pushing and pulling a wooden boat across the ice. Some had found, at a place called Starvation Cove, this boat, or a similar one, and the remains of the thirty-five men who had been dragging it. At Terror Bay the Inuit found a tent on the ice, and in it, thirty bodies. At Simpson Strait some Inuit had seen a very odd sight: the pack ice pierced by the three protruding wooden masts of a barque.

For twenty years, search parties recovered skeletons from all over the frozen sea. Franklin himself—it was learned after twelve years—had died aboard ship. Their captain dead, the ships frozen into the pack winter after winter, and supplies exhausted, the remaining officers and men had decided to walk to help. They outfitted themselves from ships' stores for the journey; their bodies were found with those supplies they had chosen to carry. Accompanying one clump of frozen bodies, for instance, which incidentally showed evidence of cannibalism, were place settings of sterling silver flatware engraved with officers' initials and family crests. A search party found, on the ice far from the ships, a letter clip, and a piece of the very backgammon board Lady Jane Franklin had given her husband as a parting gift.

Another search party found two skeletons in a boat on a sledge. These unfortunates had hauled the boat sixty-five miles. With the two skeletons were some chocolate, guns, tea, and a great deal of table silver. Many miles south of these two was another skeleton, alone. This was a frozen officer. In his pocket he had, according to Kirwan, "a parody of the sea-shanty." The skeleton was in uniform: trousers and jacket "of fine blue cloth . . . edged with silk braid, with sleeves slashed and bearing five covered buttons each. Over this uniform the dead man had worn a blue greatcoat, with a black silk neckerchief." That was the Franklin expedition.

Sir Robert Falcon Scott, who died on the Antarctic peninsula, was never able to bring himself to use dogs, let alone feed them to each other or eat them. Instead he struggled with English ponies, for whom he carried hay. Scott felt that eating dogs was inhumane. He also felt that when one reached a Pole unaided, the journey had "a fine conception" and "the conquest is more nobly and splendidly won." It is this loftiness of sentiment, this purity, this dignity and self-control, that makes Scott's farewell letters—found under his body—such moving documents.

Less moving are documents from successful polar expeditions. Their leaders relied on native technology, which, as every book ever written about the Inuit put it, was "adapted to harsh conditions."

Roald Amundsen, who returned in triumph from the South Pole, traveled Inuit-style; he made good speed using sleds and feeding dogs to dogs on a schedule. Robert E. Peary and Matthew Henson reached the North Pole in the company of four Inuit. Throughout the Peary expedition, the Inuit drove the dog teams, built igloos, and supplied seal and walrus clothing.

There is no such thing as a solitary polar explorer, fine as the conception may be.

THE PEOPLE

I have been attending Catholic Mass for only a year. Before that, the handiest church was Congregational.

Week after week I climbed the steps to that little church, entered, and took a seat with some few of my neighbors. Week after week I was moved by the pitiableness of the bare linoleum-floored sacristy no flowers could soften or cheer, by the terrible singing I so loved, by the fatigued Bible readings, the lagging emptiness and dilution of the liturgy, the horrifying vacuity of the sermon, and by the fog of dreary senselessness pervading the whole, which existed alongside, and probably caused, the wonder of the fact that we came; that we returned; that we showed up, week after week, and went through with it. Once, while we were reciting the Gloria, a farmer's wife whom I knew slightly gave me a sudden, triumphant glance.

Recently I returned to that Congregational church for an ecumenical service. A Catholic priest, together with the resident minister, served a communion of grape juice. Both the priest and the minister were professionals, old hands. They bungled with dignity and aplomb, both at ease and awed, half confident and controlled and half bewildered: "Where is it?," I could hear them whispering; "Haven't you got it?"; "I thought *you* had it!"

The priest, new to me, was in his sixties. He was tall; he wore his weariness loosely, standing upright and controlling his breath. When he knelt at the altar, and when he rose from kneeling, his knees cracked. It was a fine church music, this sound of his cracking knees.

THE LAND

Polar explorers—one gathers from their accounts—sought at the Poles something of the sublime. Simplicity and purity attracted them. They set out to perform clear tasks in uncontaminated lands. The land's austerity held them. They praised the land's spare beauty as if it were a moral or spiritual quality: "icy halls of cold sublimity," "lofty peaks perfectly covered with eternal snow." Fridtjof Nansen referred to "the great adventure of the ice, deep and pure as infinity . . . the eternal round of the universe and its eternal death." Everywhere polar prose evokes these absolutes, these ideas of "eternity" and "perfection," as if they were perfectly visible parts of the landscape.

They went, I say, partly in search of the sublime, and they found it the only way it can be found, here or there—around the edges, tucked into the corners of the days. For they were people—all of them, even the British—and, despite the purity of their conceptions, they man-hauled their humanity to the Poles along with them.

They man-hauled their frail flesh to the Poles, and encountered conditions so difficult that, for instance, it commonly took members of Scott's South Polar party several hours each morning to put on their boots. Day and night they did miserable, niggling, and often fatal battle with frostbitten toes, diarrhea, bleeding gums, hunger, weakness, mental confusion, and despair.

They man-hauled their sweet human absurdity to the Poles. When Robert E. Peary and Matthew Henson reached the North Pole in 1909, Peary planted there in the frozen ocean, according to L. P. Kirwan, the flag of the Dekes: "the colours of the Delta Kappa Epsilon Fraternity at Bowdoin College, of which Peary was an alumnus."

Polar explorers must adapt to conditions. They must adapt, on the one hand, to severe physical limitations; they must adapt, on the other hand—like the rest of us—to ordinary emotional limitations. The hard part is in finding a workable compromise. If you are Peary and have planned your every move down to the last jot and tittle, you can perhaps get away with carrying a Deke flag to the North Pole, if it will make you feel good. After eighteen years' preparation, why not feel a little good? If, on the other hand, you are an officer with the Franklin expedition and do not know what you are doing or where you are, but think you cannot eat food except from sterling silver tableware, you cannot get away with it. Wherever we go, there seems to be only one business at hand—that of finding a workable compromise between the sublimity of our ideas and the absurdity of the fact of us.

They made allowances for their emotional needs. Over-wintering expedition ships carried, in addition to sufficient fuel, equipment for the publication of weekly newspapers. The brave polar men sat cooling their heels in the middle of nowhere, reading in cold type their own and their bunkmates' gossip, in such weeklies as Parry's *Winter Chronicle* and *North Georgia Gazette*, Nansen's *Framsjaa*, or Scott's *South Polar Times* and *The Blizzard*. Polar explorers amused themselves as well with theatrical productions. If one had been frozen into the pack ice off Ross Island, near Antarctica, aboard Scott's ship *Discover*, one midwinter night in 1902, one could have seen the only performance of *Ticket of Leave*, "a screaming comedy in one act." Similarly, if, in the dead of winter, 1819, one had been a member of young William Edward Parry's expedition frozen into the pack ice in the high Arctic, one could have caught the first of a series of fortnightly plays, an uproarious success called *Miss in Her Teens*. According to Kirwan, " 'This,' Parry dryly remarked, 'afforded to the men such a fund of amusement as fully to justify the expectations we had formed of the utility of theatrical entertainments.' " And you yourself, Royal Navy Commander William Edward Parry, were you not yourself the least bit amused? Or, at the distinguished age of twenty-nine years old, did you stand on your dignity?

God does not demand that we give up our personal dignity, that we throw in our lot with random people, that we lose ourselves and turn from all that is not him. God needs nothing, asks nothing, and demands nothing, like the stars. It is a life with God that demands these things.

Experience has taught the race that if knowledge of God is the end, then these habits of life are not the means but the condition in which the means operates. You do not have to do these things, not at all. God does not, I regret to report, give a hoot. So, no, you do not have to do these things—unless, that is, you want to know God. For they work on you, not on him.

You do not *have* to sit outside in the dark. If, however, you want to look at the stars, you will find that darkness is necessary. But the stars themselves neither require nor demand it.

THE TECHNOLOGY

It is a matter for computation: How far can one walk carrying how much silver? The computer balks at the problem—too many unknowns—and puts instead its own questions: Who is this "one"? What degree of stamina may we calculate for? Under what conditions does this one propose to walk, at what latitudes? With how many companions, how much aid?

THE PEOPLE

The Mass has been building to this point, to the solemn delivery of those few hushed phrases known as the Sanctus. We have confessed, in a low, distinct murmur, our sins; we have become the people broken, and then the people made whole by our reluctant assent to the priest's proclamation of God's mercy. Now, as usual, we will, in the stillest voice, stunned, repeat the Sanctus, repeat why it is that we have come:

Holy, holy, holy Lord,
God of power and might,
heaven and earth are full of your glory . . .

It is here, if ever, that one loses oneself at sea. Here, one's eyes roll up, and the sun rolls overhead, and the floe rolls underfoot, and the scene of unrelieved ice and sea rolls over the planet's pole and over the world's rim, wide and unseen.

Now, just as we are dissolved in our privacy and

about to utter the words of the Sanctus, the lead singer of Wildflowers bursts onstage from the wings. I raise my head. He is taking enormous, enthusiastic strides, pumping his guitar's neck up and down. Drooping after him come the orange-haired country-and-western-style woman; the soprano, who, as if to shorten herself, carries her neck forward like a horse; the withdrawn boy; and then the Chinese man, holding a tambourine as if it had stuck by some defect to his fingers and he has resolved to forget about it. These array themselves in a clump downstage right. The priest is nowhere in sight.

Alas, alack, oh brother, we are going to have to sing the Sanctus.

There is, of course, nothing new about singing the Sanctus. The lead singer smiles disarmingly. He hits a chord with the flat of his hand. There is a new arrangement. The Chinese man with sudden vigor bangs the tambourine and looks at his hands, startled. They run us through the Sanctus three or four times. The words are altered a bit to fit the strong upbeat rhythm:

Heaven and earth
(Heaven and earth earth earth earth)
Are full (full full full)
Of your glory . . .

Must I join this song? May I keep only my silver? My backgammon board, I agree, is a frivolity. I relinquish it. I will leave it right here on the ice. But my silver? My family crest? One knife, one fork, one spoon, to carry beneath the glance of heaven and back? I have lugged it around for years; I am, I say, superlatively strong. Don't laugh. I am superlatively strong! Don't laugh; you'll make me laugh. The answer is no. We are singing the Sanctus, it seems, and they are passing the plate. I would rather, I think, undergo the famous dark night of the soul than encounter in church the dread hootenanny—but these purely personal preferences are of no account, and maladaptive to boot. They are passing the plate and I toss in my schooling; I toss in my rank in the Royal Navy, my erroneous and incomplete charts, my pious refusal to eat sled dogs, my watch, my keys, and my shoes. I was looking for bigger game, not little moral lessons—but who can argue with conditions?

"Heaven and earth earth earth earth," we sing. The withdrawn boy turns his head toward a man in front of me, who must be his father. Unaccountably, the enormous teenaged soprano catches my eye, exultant. A low shudder of shock crosses our floe. We have split from the pack; we have crossed the Arctic Circle, and the current has us.

THE LAND

We are clumped on an ice floe, drifting in the black polar sea. Heaven and earth are full of our terrible singing. Overhead we see a blurred, colorless brightness; at our feet we see the dulled, swift ice and recrystallized snow. The sea is black and green; a hundred thousand floes and bergs float in the water and spin and scatter in the current around us everywhere as far as we can see. The wind is cool, moist, and scented with salt.

I am wearing, I discover, the uniform of a Keystone Kop. I examine my hat: a black cardboard constable's hat with a white felt star stapled to the band above the brim. My dark Keystone Kop jacket is nicely belted, and there is a tin badge on my chest. A holster around my hips carries a popgun with a cork on a string, and a red roll of caps. My feet are bare, but I feel no cold. I am skating around on the ice, and singing, and bumping into people who, because the ice is slippery, bump into other people. "Excuse me!" I keep saying, "I beg your pardon—whoops there!"

When a crack develops in our floe and opens at my feet, I jump across it—skillfully, I think—but my jump pushes my side of the floe away, and I wind up leaping full tilt into the water. The Chinese man extends a hand to pull me out, but alas, he slips and I drag him in. We are treading water, he and I, singing, and collecting a bit of a crowd. It takes a troupe of circus clowns to get us both out. I check my uniform at once and learn that my rather flattering hat is intact, my trousers virtually unwrinkled, but my roll of caps is wet. The Chinese man is fine; we thank the clowns.

This troupe of circus clowns, I hear, is poorly paid. They are invested in bright, loose garments; a bunch of spontaneous, unskilled, oversized children, they joke and bump into people. At one end of the floe, ten of them—red, yellow, and blue—are trying to climb upon one another to make a human pyramid. It is a wonderfully funny sight, because they have put the four smallest clowns on the bottom, and the biggest, fattest clown is trying to climb to the top. The rest of the clowns are doing gymnastics; they tumble on the ice and flip cheerfully in midair. Their crucifixes fly from their ruffled necks as they flip, and hit them on their bald heads as they land. Our floe is smaller now, and we seem to have drifted into a faster bit of current. Repeatedly we ram little icebergs that rock as we hit them. Some of them tilt clear over like punching bags; they bounce back with great splashes, and water streams down their blue sides as they rise. The country-and-western-style woman is fending off some of the larger bergs with a broom. The lugs with the mustaches have found, or brought, a Frisbee, and a game is developing down the middle of our floe. Near the Frisbee game, a bunch of people including myself and some clowns are running. We fling ourselves down on the ice, shoulders first, and skid long distances like pucks.

Now the music ceases and we take our seats in the pews. A baby is going to be baptized. Overhead the sky is brightening; I do not know if this means we have drifted farther north, or all night.

THE PEOPLE

The baby's name is Oswaldo; he is a very thin baby who looks to be about one. He never utters a peep; he looks grim, stiff as a planked shad. His parents—his father carrying him—and his godparents, the priest, and the two acolytes, are standing on the ice between the first row of pews and the linoleum-floored sacristy. I am resting my bare feet on the velvet prie-dieu—to keep those feet from playing on the ice during the ceremony.

Oswaldo is half-Filipino. His mother is Filipina. She has a wide mouth with much lipstick, and wide eyes; she wears a tight black skirt and stiletto heels. The father looks like Ozzie Nelson. He has marcelled yellow hair, a bland, meek face, and a big, meek nose. He is wearing a brown leather flight jacket. The godparents are both Filipinos, one of whom, in a pastel denim jumpsuit, keeps mugging for the camera another family member is shooting from the aisle.

The baby has a little red scar below one eye. He is wearing a long white lace baptismal gown, blue tennis shoes with white rubber toes, and red socks. The priest anoints the baby's head with oil. He addresses to the parents several articles of faith: "Do you believe in God, the Father Almighty, creator of Heaven and earth?"

"Yes, we believe."

The priest repeats a gesture he says was Christ's, explaining that it symbolically opens the infant's five senses to the knowledge of God. Uttering a formal prayer, he lays his hand loosely over Oswaldo's face and touches in rapid succession his eyes, ears, nose, and mouth. The baby blinks. The priest, whose voice is sometimes lost in the ruff at his neck, is formal and gentle in his bearing; he knows the kid is cute, but he is not going to sentimentalize the sacrament.

Since our floe spins, we in the pews see the broken floes and tilting bergs, the clogged, calm polar sea, and the variously lighted sky and water's rim shift and revolve enormously behind the group standing around the baby. Once I see a yellowish polar bear erupt out of the water as smoothly as if climbing were falling. I see the bear splash and flow into a distant floeberg that tilts out of sight.

Now the acolytes bring a pitcher, a basin, and a linen towel. The father tilts the rigid baby over the basin; the priest pours water from the pitcher over the baby's scalp; the mother sops the baby with the linen towel and wraps it over his head, so that he looks, proudly, as though he has just been made a swami.

To conclude, the priest brings out a candle, for the purpose, I think, of pledging everybody to Christian fellowship with Oswaldo. Actually, I do not know what it is for; I am not listening. I am watching the hands at the candlestick. Each of the principals wraps a hand around the brass candlestick: the two acolytes with their small, pale hands at its base; the two families—Oswaldo's and his godparents'—with their varicolored hands in a row, and the priest at the top, as though he has just won the bat toss at baseball. The baby rides high in his father's arms, pointing his heels in his tennis shoes, silent, wanting down. His father holds him firmly with one hand and holds the candlestick beside his wife's hand with the other. The priest and the seated members of Wildflowers start clapping then—a round of applause for everybody here on the ice!—so we clap.

II

MONTHS HAVE PASSED; years have passed. Whatever ground gained has slipped away. New obstacles arise, and faintness of heart, and dread.

THE LAND

Polar explorers commonly die of hypothermia, starvation, scurvy, or dysentery; less commonly they contract typhoid fever (as Stefasson did), vitamin A poisoning from polar bear liver, or carbon monoxide poisoning from incomplete combustion inside tents sealed with snow. Very commonly, as a prelude to these deaths, polar explorers lose the use of their feet; their frozen toes detach when they remove their socks.

Particularly vivid was the death of a certain Mr. Joseph Green, the astronomer on Sir James Cook's first voyage to high latitudes. He took sick aboard ship. One night "in a fit of the phrensy," as a contemporary newspaper reported, he rose from his bunk and "put his legs out of the portholes, which was the occasion of his death." Vitus Bering, shipwrecked in 1740 on Bering Island, was found years later preserved in snow. An autopsy showed he had many lice, as well as scurvy, and died of a "rectal fistula which forced gas gangrene into his tissues."

The bodies of various members of the Sir John Franklin expedition of 1845 were found over the course of twenty years, by thirty search expeditions, in assorted bizarre postures scattered over the ice of Victoria Strait, Beechey Island, and King William Island.

Sir Robert Falcon Scott reached the South Pole on January 17, 1912, only to discover a flag that Roald Amundsen had planted there a month earlier. Scott's body, and the bodies of two of his companions, turned up on the Ross Ice Shelf eleven miles south of one of their own supply depots. The bodies were in sleeping bags. Scott's journals and farewell letters, found under his body, indicate that the other two had died first. His torso was well out of his sleeping bag, and, to get the dying over with, he had opened wide the collar of his parka, exposing his skin.

Of the United States Greely expedition to the North Pole, all men died but six. Greely himself, one of the six survivors, was found "on his hands and knees with long hair in pigtails." Of the United States De Long expedition to the North Pole in the *Jeannette*, all men died but two. Of the *Jeannette* herself and her equipment, nothing was found until three years after she sank, when, on a beach on the other side of the polar basin, a Greenlander discovered a pair of yellow oilskin breeches stamped JEANNETTE.

Never found were the bodies of Henry Hudson, his young son, and four men, whom mutineers in 1610 had lowered from their ship in a dinghy, in Hudson's Bay, without food or equipment. Never found were the bodies of Sir John Franklin, or of Amundsen and seventeen other men who set out for the Arctic in search of a disastrous Italian expedition, or the bodies of Scott's men Evans and Oates. Never found were most of the drowned crew of the United States ship *Polaris* or the body of her commander, who died sledging on the ice.

THE PEOPLE

Why do we people in churches seem like cheerful, brainless tourists on a packaged tour of the Absolute? The tourists are having coffee and doughnuts on Deck C. Presumably someone is minding the ship, correcting the course, avoiding icebergs and shoals, fueling the engines, watching the radar screen, noting weather reports radioed in from shore. No one would dream of asking the tourists to do these things. Alas, among the tourists on Deck C, drinking coffee and eating doughnuts, we find the captain, all the ship's officers, and all the ship's crew. The officers chat; they swear; they wink a bit at slightly raw jokes, just like regular people. The crew members have funny accents. The wind seems to be picking up.

On the whole, I do not find Christians outside of the catacombs sufficiently sensible of conditions. Does anyone have the foggiest idea what sort of power we so blithely invoke? Or, as I suspect, does no one believe a word of it? The churches are children playing on the floor with their chemistry sets, mixing up a batch of TNT to kill a Sunday morning. It is madness to wear ladies' straw hats and velvet hats to church; we should all be wearing crash helmets. Ushers should issue life preservers and signal flares; they should lash us to our pews. For the sleeping god may wake someday and take offense or the waking god may draw us out to whence we can never return.

The eighteenth-century Hasidic Jews had more sense, and more belief. One Hasidic slaughterer, whose work required invoking the Lord, bade a tearful farewell to his wife and children each morning before he set out for the slaughterhouse. He felt, every morning, that he might never see any of them again. For every day, as he stood with his knife in his hand, the words of his prayer carried him into danger. Once he called on God, God could so easily destroy him after all, before he had time to utter the words, "Have mercy."

Another Hasid, a rabbi, refused to promise a friend to visit him the next day: "How can you ask me to make such a promise? This evening I must pray and recite

'Hear, O Israel.' When I say these words, my soul goes out to the utmost rim of life. . . . Perhaps I shall not die this time either, but how can I now promise to do something at a time after the prayer?"

ASSORTED WILDLIFE

Insects

I like insects for their stupidity. A paper wasp—*Polistes*—is fumbling at the stained-glass window on my right. I saw the same sight in the same spot last Sunday: Pssst! Idiot! Sweetheart! Go around by the door! I hope we seem as endearingly stupid to God—bumbling down into lamps, running half-wit across the floor, banging for days at the hinge of an opened door. I hope so. It does not seem likely.

Penguins

Antarctic penguins, according to visitors, are . . . adorable. They are tame! They are funny! Tourists in Antarctica are mostly women of a certain age. They step from the cruise ship's rubber Zodiacs wearing bright ship's-issue parkas; they stalk around on the gravel and squint into the ice glare; they exclaim over the penguins, whom they find tame, funny, and adorable; they take snapshots of one another with the penguins, and look around cheerfully for something else to look around at.

The penguins are adorable, and the wasp at the stained-glass window is adorable, because in each case their impersonations of human dignity so evidently fail. What are the chances that God finds our own failed impersonation of human dignity adorable? Or is he fooled? What odds do you give me?

III

THE LAND

SEVERAL YEARS AGO I visited the high Arctic and saw it: the Arctic Ocean, the Beaufort Sea. The place was Barter Island, inside the Arctic Circle, in the Alaskan Arctic north of the North Slope. I stood on the island's ocean shore and saw what there was to see: a pile of colorless stripes. Through binoculars I could see a bigger pile of colorless stripes.

It seemed reasonable to call the colorless stripe overhead "sky," and reasonable to call the colorless stripe at my feet "ice," for I could see where it began. I could distinguish, that is, my shoes and the black gravel shore and the nearby frozen ice the wind had smashed ashore.

It was this mess of ice—ice breccia, pressure ridges, and standing floes, ice sheets upright, tilted, frozen together and jammed—that extended out to the horizon. No matter how hard I blinked, I could not put a name to any of the other stripes. Which was the horizon? Was I seeing land, or water, or their reflections in low clouds? Was I seeing the famous "water sky," the "frost smoke," or the "ice blink"?

In his old age, James McNeill Whistler used to walk down to the Atlantic shore carrying his paints. Day after day, he painted broad, blurred washes representing sky, water, and shore, three blurry light-filled stripes. These are late Whistlers; I like them very much, and in the high Arctic I thought of them, for I seemed to be standing in one of them. If I loosed my eyes from my shoes, the gravel at my feet, or the chaos of ice at the shore, I saw what newborn babies must see: nothing but senseless variations of light. The world was a color-field painting wrapped round me at an unknown distance. I hesitated to take a step.

There was, in short, no recognizable three-dimensional space in the Arctic. There was also no time. The sun never set, but neither did it appear. The dim round-the-clock light changed haphazardly when the lid of cloud thickened or thinned. Circumstances made the eating of meals random and even impossible. I slept when I was tired. When I woke I walked out into the colorless stripes and the revolving winds, where atmosphere mingled with distance, and where land, ice, and light blurred into a dreamy, freezing vapor that, lacking anything else to do with the stuff, I breathed. Now and then a white bird materialized out of the vapor and screamed. It was, in short, what one might, searching for words, call beautiful land; it was more beautiful still when the sky cleared and the ice shone in the dark water.

THE TECHNOLOGY

It is for the Pole of Relative Inaccessibility I search and have been searching in the mountains and along the seacoasts for years. The aim of this expedition is, as Pope Gregory put it in his time, "To attain to somewhat of the unencompassed light, by stealth, and scantily." How often have I mounted this same expedition, has my absurd barque set out, half-caulked, for the Pole?

THE LAND

"These incidents are true," I read in an 1880 British history of Arctic exploration. "These incidents are true—the storm, the drifting ice-raft, the falling berg, the sinking ship, the breaking up of the great frozen floe: these scenes are real—the vast plains of ice, the ridged hummocks, the bird-thronged cliff, the far-stretching glacier."

Polar exploration is no longer the fashion it was during the time of the Franklin expedition, when beachgoers at Brighton thronged to panoramas of Arctic wastes painted in shop windows, and when many thousands of Londoners jammed the Vauxhall pleasure gardens to see a diorama of polar seas. Our attention is elsewhere now, but the light-soaked land still exists: I have seen it.

THE TECHNOLOGY

In the nineteenth century, a man deduced Antarctica.

During that time, no one on earth knew for certain whether there was any austral land mass at all, although the American Charles Wilkes claimed to have seen it. Some geographers and explorers speculated that there was no land, only a frozen Antarctic Ocean; others posited two large islands in the vicinity of the Pole. That there is a South Polar continent was not in fact settled until 1935.

In 1893, one John Murray presented to the Royal Geographic Society a deduction of the Antarctic continent. His expedition's ship, the *Challenger,* had never come within sight of any such continent. His deduction proceeded entirely from dredgings and soundings. In his presentation he posited a large, single continent, a speculative map of which he furnished. He described accurately the unknown continent's topology: its central plateau with its permanent high-pressure system, its enormous glacier facing the Southern Ocean, its volcanic ranges at one coast, and at another coast, its lowland ranges and hills. He was correct.

Deduction, then, is possible. There are many techniques for the exploration of high latitudes. There is, for example, such a thing as a drift expedition.

When a pair of yellow oilskin breeches belonging to the lost crew of the *Jeannette* turned up after three years in Greenland, having been lost north of central Russia, Norwegian explorer Fridtjof Nansen took note. On the basis of these breeches' travels he plotted the probable direction of the current in the polar basin. Then he mounted a drift expedition: In 1893 he drove his ship, the *Fram,* deliberately into the pack ice and settled in to wait while the current moved north and, he hoped, across the Pole. For almost two years, he and a crew of twelve lived aboard ship as the frozen ocean carried them. "I long to return to life," Nansen wrote in his diary, ". . . the years are passing here. . . . Oh! At times this inactivity crushes one's very soul; one's life seems as dark as the winter night outside; there is sunlight upon no other part of it except the past and the far, far distant future. I feel as if I must break through this deadness."

The current did not carry them over the Pole, so Nansen and one companion set out one spring with dog sledges and kayaks to reach the Pole on foot. Conditions were too rough on the ice, however, so after reaching a record north latitude, the two turned south toward land, wintering together finally in a stone hut on Franz Josef Land and living on polar bear meat. The following spring, after almost three years, they returned to civilization.

Nansen's was the first of several drift expeditions. During World War I, members of a Canadian Arctic expedition, camping on an ice floe seven miles by fifteen miles, drifted in the Beaufort Sea over four hundred miles in the course of six months. In 1937, an airplane deposited a Soviet drift expedition on an ice floe near the North Pole. These four Soviet scientists drifted for nine months while their floe, colliding with grounded ice, repeatedly split into ever-smaller pieces.

THE LAND

I have, I say, set out again.

The days tumble with meanings. The corners heap up with poetry; whole unfilled systems litter the ice.

THE TECHNOLOGY

A certain Lieutenant Maxwell, a member of Vitus Bering's second polar expedition, wrote, "You never feel safe when you have to navigate in waters which are completely blank."

Cartographers call blank spaces on a map "sleeping beauties."

On our charts I see the symbol for shoals and beside it the letters "P.D." My neighbor in the pew, a lug with a mustache who has experience of navigational charts and who knows how to take a celestial fix, tells me that the initials stand for "Position Doubtful."

THE LAND

To learn the precise location of a Pole, once you're near one, choose a clear, dark night to begin. Locate by ordinary navigation the Pole's position within an area of several square yards. Then arrange on the ice in that area a series of loaded cameras. Aim the cameras at the sky's zenith; leave their shutters open. Develop the film. The film from that camera located precisely at the Pole will show the night's revolving stars as perfectly circular concentric rings.

THE TECHNOLOGY

I have a taste for solitude, and silence, and for what Plotinus called "the flight of the alone to the Alone." I have a taste for solitude. Sir John Franklin had, apparently, a taste for backgammon. Is either of these appropriate to conditions?

You quit your house and country, quit your ship, and quit your companions in the tent, saying, "I am just going outside and may be some time." The light on the far side of the blizzard lures you. You walk, and one day you enter the spread heart of silence, where lands dissolve and seas become vapor and ices sublime under unknown stars. This is the end of the Via Negativa, the lightless edge where the slopes of knowledge dwindle, and love for its own sake, lacking an object, begins.

THE LAND

I have put on silence and waiting. I have quit my ship and set out on foot over the polar ice. I carry chronometer and sextant, tent, stove and fuel, meat and fat. For water I melt the pack ice in hatchet-hacked chips; frozen salt water is fresh. I sleep when I can walk no longer. I walk on a compass bearing toward geographical north.

I walk in emptiness; I hear my breath. I see in my

hand a compass, see the ice so wide it arcs, see the planet's peak curving and its low atmosphere held fast on the dive. The years are passing here. I am walking, light as any handful of aurora; I am light as sails, a pile of colorless stripes; I cry "heaven and earth are indistinguishable!" and the current underfoot carries me and I walk.

The blizzard is like a curtain; I enter it. The blown snow heaps in my eyes. There is nothing to see or to know. For weeks I wait in the tent, adrift and emptied, while the storm unwinds. One day it is over, and I pick up my tent and walk. The storm has scoured the air; the clouds have lifted. The sun rolls round the sky like a fish in a round bowl, like a pebble rolled in a tub, like a swimmer, or a melody flung and repeating, repeating enormously overhead on all sides.

My name is Silence. Silence is my bivouac, and my supper sipped from bowls. I robe myself each morning in loose strings of stones. My eyes are stones; a chip from the pack ice fills my mouth. My skull is a polar basin; my brain pan grows glaciers, and icebergs, and grease ice, and floes. The years are passing here.

Far ahead is open water. I do not know what season it is, know how long I have walked into the silence like a tunnel widening before me, into the horizon's spread arms that widen like seas. I walk to the pack ice edge, to the rim which calves its floes into the black and green water; I stand at the edge and look ahead. As far as I can see, a scurf of candle ice on the water's skin scratches the sea and crumbles whenever a lump of ice or snow floats through it. The floes are thick in the water, some of them large as land. By my side is passing a flat pan of floe from which someone extends an oar. I hold the oar's blade, jump, and land on the long floe.

No one speaks. Here, at the bow of the floe, the bright clowns have staked themselves to the ice. With tent stakes and ropes they have lashed their wrists and ankles to the floe on which they lie stretched and silent, faceup. Among the clowns, and similarly staked, are many boys and girls, some women, and a few men from various countries. One of the men is Nansen, the Norwegian explorer who drifted. One of the women repeatedly opens and closes her fists. One of the clowns has opened his neck ruffle, exposing his skin. For many hours I pass among these staked people, intending to return later and take my place.

Farther along I see that the tall priest is here, the priest who served grape-juice communion at an ecumenical service many years ago, in another country. He is old. Alone on a wind-streaked patch of snow he kneels, then stands; kneels, then stands; and kneels again. Not far from him, at the floe's side, sitting on a packing crate, is the deducer John Murray. He lowers a plumb bob overboard and pays out the line. He is wearing the antique fur hat of a Doctor of Reason, such as Erasmus wears in his portrait; it is understood that were he ever to return and present his finding, he would be ridiculed, if only for his hat. Scott's Captain Oates is here; he has no feet. It is he who stepped outside his tent, to save his friends. Now, on his dignity, he stands and mans the sheet of a square linen sail; he has stepped the wooden mast on a hillock amidships.

From the floe's stern I think I hear music; I set out, but it takes me several sleeps to get there. I am no longer using the tent. Each time I wake, I study the floe and the ocean horizon for signs—signs of the pack ice we left behind, of open water, or land, or any sort of weather. But nothing changes; there is only the green sea and the floating ice, and the black sea in the distance, speckled by bergs, and a steady wind astern which smells of unknown mineral salts, some ocean floor.

At last I reach the floe's broad stern, its enormous trailing coast, its throngs, its many cooking fires. There are children carrying babies, and men and women painting their skins and trying to catch their reflections in the water to leeward. Near the water's edge is a wooden upright piano, and a bench with a telephone book on it. A woman is sitting on the telephone book and banging out the Sanctus on the keys. The wind is picking up. I am singing at the top of my lungs, for a lark.

Many clowns are here; one of them is passing out Girl Scout cookies, all of which are stuck together. Recently, I learn, Sir John Franklin and crew have boarded this floe, and so have the crews of the lost *Polaris* and the *Jeannette*. The men, whose antique uniforms are causing envious glances, are hungry. Some of them start rough-housing with the rascally acolyte. One crewman carries the boy on his back along the shore to the piano, where he abandons him for a clump of cookies and a seat on the bench beside the short pianist, whose bare feet, perhaps on account of the telephone book, cannot reach the pedals. She starts playing "The Sound of Music."

"You know any Bach?" I say to the lady at the piano, whose legs seem to be highly involved with those of the hungry crewman; "You know any Mozart? Or maybe 'How Great Thou Art'?"

A skeletal officer wearing a black silk neckerchief has located Admiral Peary, recognizable from afar by the curious flag he holds. Peary and the officer together are planning a talent show with skits. When they approach me, I volunteer to sing "Antonio Spangonio, That Bum Toreador" and/or to read a piece of short fiction; they say they will get back to me.

Christ, under the illusion that we are all penguins, is crouched down posing for snapshots. He crouches, in his robe, between the lead singer of Wildflowers, who is joyfully trying to determine the best angle at which to hold his guitar for the camera, and the farmer's wife, who keeps her eyes on her painted toenails until the Filipino godfather says "Cheese." The country-and-western woman, singing, succeeds in pressing a cookie upon the baby Oswaldo. The baby Oswaldo is standing in his lace gown and blue tennis shoes in the center of a circle of explorers, confounding them all.

In my hand I discover a tambourine. Far ahead, out on the brittle horizon, I see icebergs among the floes. I see tabular bergs and floebergs and dark cracks in the water between them. Low overhead on the underside of the thickening cloud cover are dark colorless stripes reflecting pools of open water in the distance. I am banging on the tambourine, and singing whatever the piano player plays; now it's "On Top of Old Smokey." I am banging the tambourine and belting the songs so loudly that people are edging away. But how can any of us tone it down? For we are nearing the Pole.
